HANDS, THE ACHILLES' HEEL

HANDS, THE ACHILLES' HEEL

THE UNDISCLOSED LOGIC OF HUMAN BEHAVIOUR

TOWARDS AN UNDERSTANDING OF AUTONOMY, HETERONOMY AND HUMAN FREEDOM

PETER FFITCH

Matador
9 Priory Business Park,
Wistow Road, Kibworth Beauchamp,
Leicestershire. LE8 0RX
Tel: 0116 279 2299
Email: books@troubador.co.uk
Web: www.troubador.co.uk/matador
Twitter: @matadorbooks

ISBN 978 1788033 015

British Library Cataloguing in Publication Data.
A catalogue record for this book is available from the British Library.

Printed and Bound in the UK by 4Edge limited
Typeset in 11pt Adobe Garamond Pro by Troubador Publishing Ltd, Leicester, UK

Matador is an imprint of Troubador Publishing Ltd

To the countless millions who have lost their autonomy
at the hands of others.

Cave painting, Kolo, Tanzania, thought to be 23,000 years old. This image of abduction is one of the earliest depictions of hominin motor-control.

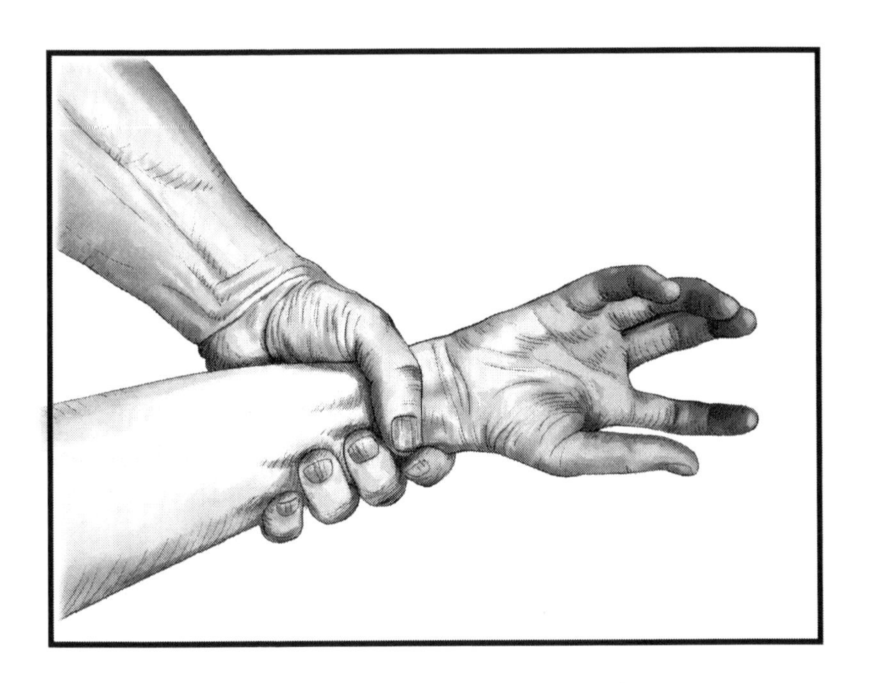

CONTENTS

PREFACE

We do not have a convincing explanation of human behaviour. It was the existence of this gap in our understanding that first attracted my attention when I was quite young. Not that there was any realistic possibility of making inroads into the problem for, at that time, the whole subject was recognized as being far too complex to resolve. However, the lack of an established answer did create an opportunity, for I was free to explore, to see where my own musings may lead.

Having some rudimentary understanding of all this, by the time I had to earn a living I turned to a career that would not constrain my independent thinking. I trained in agriculture, purchased a small farm and became a livestock farmer. Farming opened a range of animal behaviour for observation, including the domestic livestock on the farm, the more domesticated pets living in the farmhouse and the local wild animals that lived in the fields, woods and rivers of the surrounding countryside. Over the next forty years I observed these animals not appreciating, at first, that a deep awareness of animal behaviour would eventually lead me to a radical reconceptualization of human behaviour.

Livestock farmers have an interesting relationship with their animals since they are totally responsible for their welfare; they not only have to organize the animals' nutritional needs and health care but also have to supervise the mating of all the female animals on the farm, which in my case included mares, cows, ewes, sows, bitches, goats, hens and geese. I became skilled in managing the vagaries of thousands of these pairings and this is where I learnt how essential the rhythm of the oestrous/anoestrous cycle was to the safety of the female animal.

In contrast to the controls and restrictions of farming I also witnessed, as I strolled in the countryside, the underlying balance and harmony of the wild animals' autonomous behaviour. As their composure worked upon my sensibilities I absorbed the unspoken flow of animal life without fully realizing the extent of their freedom of movement until, on a cold wet February afternoon in 1973, it became pivotal to my understanding of humankind and the reason for our deep-seated malaise.

INTRODUCTION

Hands, the Achilles' Heel is a new theory of human evolution. It answers the long-standing problem that arose out of Darwin's *Origin of Species* – how is it that, having evolved from the animals, we now consider ourselves to be so 'exceptional' that we no longer see ourselves as animals? It is about an undisclosed logic of human behaviour that in theory could have been understood a long time ago since the evidence upon which its conclusions are based are everyday experiences that have not changed for hundreds of thousands of years.

More than 250 years ago thinkers such as David Hume (1711–76) and Jean-Jacques Rousseau (1712–78) independently set out to establish 'a science of human nature'; however, they failed to achieve their goal and since that time progress on the project has been strangely disappointing. Sigmund Freud (1856–1939) was the last influential thinker to suggest a comprehensive new theory, but his ideas failed the test of scientific credence and no other thinker has taken up the challenge.

A deep understanding of the origins of human nature should, by now, have emerged in the space bounded by other disciplines such as, Human Evolutionary Biology, Animal Behaviour, Primatology, Anthropology, Philosophy, Psychology and Psychoanalysis. Over the years, there have been glimmers of light but the reasons for our unique behaviour and psychology have remained as elusive as ever.

There is also an additional difficulty: any enquiry concerning human nature needs to be prepared to do battle with what Giorgio Agamben (and others) call the 'Anthropological Machine' which, at every opportunity, trumpets human exceptionalism at the expense of the so-called distinctly 'inferior' animal, leaving a trail of misinformation in its wake. (Species differ but one cannot be classed as 'superior' to another. Individuals exploit their environmental niche and if their offspring survive then that species is successful).

If we are to achieve a more honest understanding of ourselves, and our place in nature, we will have to be prepared not only to see ourselves as animals but to also see that animals (across all species) have a similar enviable, equitable mental stance towards the world that humans are unable to share because we are so tense and anxious. We need to discover how a species that was originally 'part of nature' could, in an undisclosed way, come to find that it had become human and somehow 'outside of nature'. We cannot know who we are until we understand the selection pressures that have moulded our own evolution.

The standard reply is that we evolved to become more intelligent, but that does not explain why we, amongst the myriad of other animals (given that many lead a precarious existence) are the only species that needed to become more intelligent in order to survive.

The human breakout from the animal world cries out for an explanation. How did it happen, why did it happen?

There had to be a seismic shift in the evolutionary selection pressures for it to occur, yet mankind's response so far (which is telling in itself) is that, as beneficiaries, we should simply take advantage of the situation and bask in our exceptionalism.

Asking how we have come to find ourselves outside of nature is essentially the same question as the one addressed by psychoanalysts when they ask, "How can there be an animal (albeit a human animal) that represses itself?" Or by religious communities, when they ask, "Whence evil?" Or by scientists who seek to find the evolutionary selection pressures that led to the divergence of the early hominins from the apes. A single answer, therefore, has the potential of ultimately uniting these disparate communities because, at base, they are all addressing the same problem.

Given that we have evolved from the animals, by means of the standard evolutionary principles of selection, it is a puzzle to find that there are several major differences between animals and humans that defy an explanation. These are not the typical variations between animal species such as having either hooves, pads or paws, they are differences unique to human anatomy and behaviour and they include such things as bipedalism, hairlessness, breasts on non-lactating females, loss of the ability to defend anoestrus, the enlarged brain and the development of speech and language, all of which, at present, have no accepted explanation – that is, we do not understand the evolutionary pressure(s) that gave rise to their selection. (The latest comprehensive guide to these mysteries and the problems associated with their current explanations can be found in *Homo Mysterious* by Professor David Barash, Oxford, 2012.)

I discuss these problems in the following chapters and suggest that they are resolved by understanding that the evolutionary driving force associated with the rise of the hominins, and eventually *Homo sapiens,* is the motor-control of one hominin by another. That is the vulnerability of one hominin to having its behaviour directed by another.

I suggest (and here is the central thesis of this book) that the selection pressure that created the remarkable difference between animals and humans arose from the use of the hominin hand. At some stage in the last six million years, a group of tree-living primates became adapted to life on the ground. This had unexpected consequences, for it released the primate hand from the restraint of its original arboreal function, that of holding on to the branches in the treetops. Thus, as soon as the terrestrial lifestyle was adopted, hands were free to engage in other activities and at this point most commentators focus upon our stunning ability to use our hands for such things as making tools, weapons and building houses. In so doing they fail to see that a pair of hands, each with an opposable thumb, positioned at the end of an arm that has a flexible shoulder and elbow (designed for brachiation), are (alas) the perfect appendage with which to grab, grapple, hold and direct the behaviour of others.

This meant that the hominin male was able to catch and hold the female against her will. The consequences of this easily accomplished (almost casual) act have been wide-ranging and catastrophic (especially for the female). Being able to grasp and hold creates a unique ability to control the behaviour of other members of one's own species and this, I suggest, is the undisclosed driving force of human evolution. I have called this ability motor-control.

Everything that follows explains or expands upon the content of the last two paragraphs, for this book attempts to gain an initial understanding of the consequences that flow from having this transforming ability to grip and hold a conspecific and it sets in motion a reassessment of what it means to be human. It is the start of a journey in which we will come to understand that we have all lost the ability to function autonomously. Individuals of all species are born with a system of autonomous governance that has been selected because it best ensures their survival. When the hominins gained the ability to motor-control fellow members of their own species they were able to overwhelm and hijack the autonomous governance of their victims and impose their own heteronomous controls. (Today we fall victim to an unlimited number of controllers including trespassing males, parents, partners, friends, enemies, teachers, religious elders, trainers, employers and the state).

We are pressured, obliged or forced into enacting the controls, directions and desires of others; this is heteronomous behaviour because it is sourced in someone else's mind. No animal comes anywhere near to having the demeaning experience of being in thrall to another member of its own species, for they take the greatest care (up until they are close to the point of death) to protect and vigorously maintain their own autonomous governance.

Heteronomous control has proved to be a strangely difficult problem with which to come to terms, for despite (or because of) the fact that its disruptive effects influence our behaviour every moment of every day we take it as normal and so we are mostly 'unaware' of its operation.

It appears that it has been hidden in plain sight because we have been unable to isolate the problem and face up to the extraordinary damage that it does to the individual and all human relationships. However, there is a complication that needs to be understood from the beginning. Because we live in a world of motor-control we are constantly seeking defences to those controls. Some of the most effective defences involve 'cooperation' to outnumber any potential controllers. Yet, this type of cooperation is itself dependent upon the heteronomous control and organization of family, friends or group leaders. So, some controls have a beneficial aspect but only in a situation where they create defences to the more dangerous forms of overt physical motor-control such as rape, abduction, torture, slavery, servitude and bondage. Thus, heteronomous cooperation has become a useful defence in a dire situation but the fact remains that heteronomous mental collaboration does not exist in the animal world because animals virtually never experience overt motor-control, and so they never lose their autonomy at the hands of a conspecific.

We humans are the victims of a real-life tragedy (there are

no words that can adequately describe the consequences of this situation): we have become ensnared because the hominins are the only species that have used their hands to grip, hold and control others. In addition, because hands are physically relatively gentle, compared to the use of claws, jaws and teeth, the victim (the controlee) remains alive and remembers and reacts to the fear and loss of autonomy for the rest of her, or his, life. By overriding autonomy in this way, hands have created a never-ending tsunami of harsh psychological suffering that is too large to quantify.

It is time to see *Homo sapiens* (and all the other hominins) for what they have always been throughout the whole of hominin evolution – as a species that easily becomes victim to motor-control and heteronomy. Interestingly, owning up to this tragedy also gives rise to hope, for if we now gain an understanding of the predicament we are in, it will open a science-based possibility that this difficult era could come to an end. For, as we shall see, the uniquely human suffering we experience has arisen solely because motor-control overrides our underlying autonomous governance and this trespass necessarily disrupts the equanimity of the mind, leaving the controlee permanently anxious and disturbed, searching for safety, meaning and purpose in ways that are outside of the animal experience.

The thoughts set down in this book amount to a far-reaching reassessment of the human condition and they should be understood as the first tentative steps towards our understanding of autonomy, heteronomy and the science of human freedom. They suggest an answer to the long-standing puzzle of human behaviour which, in turn, opens a vast new territory of enquiry, so there is no pretence that this is a definitive account.

PROLOGUE

The key to understanding human nature lies in an area that has remained stubbornly resistant to any real exploration since Jean-Jacques Rousseau started to map its boundary in *A Discourse on Inequality*, when he said, "… it is no light enterprise to separate that which is original from that which is artificial in man's present nature, and attain a solid knowledge of a state that no longer exists, … yet of which it is necessary to have sound ideas if we are to judge our present state satisfactorily." (Everyman's University Library, Jean-Jacques Rousseau, *The Social Contract and Discourses*.)

Rousseau eventually abandoned this task for he had to admit that he could not give a specific account of why the transition from the 'state of nature' to 'civil society' had occurred. Hence the famous opening paragraph of *The Social Contract*, "Man is born free; and everywhere he is in chains… How did this change come about? I do not know. What can make it legitimate? That question I think I can answer." However, Rousseau went on to suggest that philosophers had not gone far enough back into nature, and on the frontispiece of the *Discourse on Inequality*, he quoted Aristotle, "It is not in depraved beings, but in those who act in accordance with nature that we must seek what is natural." (Aristotle's *Politics*, bk. 2.)

To seek what is natural in the field of human behaviour, and thereby understand the transition from freedom to chains, and the subsequent adoption of the social contract, it has been necessary to go back as far as evolutionary philosophy will allow – to the time of the protohominins, and re-ask Rousseau's original question in the language of the twenty-first century – what were the evolutionary pressures that drove the selection of the hominins (Endnote 1) forcing them to diverge from the apes and then to gradually evolve into *Homo sapiens* occupying a place in nature that is separate from that of all other animals?

The social contract is central to this discussion. Social contract theory overlays two specialities – political history and human evolutionary biology – and while the early theorists such as Grotius, Hobbes and Rousseau had a sound understanding of political history, their pre-Darwinian knowledge of natural history was necessarily rudimentary. Since this time, evolutionary biologists have made some progress and they now have a reasonably clear idea of the physical stages involved in human evolution but so far, the philosophers and neuroscientists have failed to match them with an acceptable understanding of the concomitant mental changes that must also have occurred, hence we are still floundering. As the ideas of the early contract theorists still form the basis of many of the ways we think about ourselves it will be useful to address some of the discrepancies that have emerged between them and the new scientific models.

Obviously, the objective of a social contract is to achieve social order, so we first need to differentiate between three types of social order:

1. The social order of all animals, excluding the hominins, but including the protohominin apes that existed before 5–7 million years ago.
2. The social order of the early hominins, which include primitive mankind or pre-Stone-Age man, as well as Stone-Age mankind or 'savage mankind', who existed

from the ape/hominin transition to the early years of civilization.

3. The social order of Civilized Mankind, from about 10,000 BC up to the present day.

Starting with the social order of all animals, I am not here concerned with the functional details that vary from species to species, such as dominance hierarchies or patterns of social cohesion, but that animals achieve their species-specific social order without recourse to top-down culturally imposed social laws and contracts (as the sage said, "Bears have no government and I do not see they have need of one.")

Their innate social orders, which are maintained without the benefit of a human type of intelligence, are stable, collapsing only at times of extreme climatic stress, pollution or disease. This social order is established because evolution can do no other than always select behaviour that facilitates the female's ability to successfully produce and rear their young.

From this understanding of natural selection, together with the new information coming from ethological studies, it is now accepted that the social relationships of animals of the same species (intraspecific) are not 'red in tooth and claw', on the contrary they have a high degree of social order (and it should not be confused with predatory behaviour between species – interspecific – which is highly disruptive to the prey).

This intraspecific order that was best fitted for survival has been selected over millions of years of evolution and degrees of order do not (cannot) come any higher than this. Yet there is still a common belief that mankind has passed from the turmoil and disorder in the state of nature, to the peace, cooperation, and order of the civil society. The hominins did not create a new and commendable boundary between their law-abiding selves and the pre-existing vicious, chaotic, threatening, savage norm of nature

– quite the reverse. Hominins evolved from the animal pattern (with its relatively peaceful intraspecific behaviour) to become a new species struggling to restore social cohesion because their ancestors had developed the ability to override the autonomy of their conspecifics.

Several points arise from the 'red in tooth and claw' misapprehension. When making comparisons between animals and mankind we have to watch for lapses in reasoning that state that animals are 'beasts' in the sense that they exhibit uncontrolled or irrational aggression. As noted, the brutishness that we associate with animals is generally interspecific predation not intraspecific violence. And when intraspecific attacks do occur in the animal world, for example in dominance battles, the vanquished is usually able to retreat before receiving fatal injuries.

Thus, it is rare for wild animals to experience the serious violation of having their autonomy overridden in intraspecific attacks. In this sense, the hominins experience far more intraspecific violence than animals and they do so even though they have the benefit of a social contract which presumably provides *Homo sapiens* with a considerable degree of protection.

Even with the protection of a social contract, humans experience other forms of intraspecific coercion that are never experienced by animals, including torture, slavery, genocide, imprisonment, and the human female is under a constant threat of rape and sexual harassment, which is virtually non-existent for her animal counterparts because they are always able to establish anoestrus which confirms they are completely safe from male harassment.

Now these are very telling differences; like any other mammalian species, *Homo sapiens* have been selected by evolution, so we need to determine just how and when these differences came about, and to which pressures they are responses, for violence, particularly violence directed towards the breeding female (who in evolutionary terms is a prized resource) is very difficult to explain.

The point I wish to establish here is that extant primates (and by inference also the extinct 'animal' protohominins) have social relationships that from time to time have serious flashpoints, but these flashpoints do not permanently override the autonomy of the individual and they do not disrupt the organization of society to the extent that the species 'fall out of harmony with nature, requiring its members to enter into a social contract to re-establish a tenable social order.

It has always been accepted that mankind is a special case in the evolutionary order of things, or as Spinoza conceived it – "situated in nature as a dominion within a dominion". I wish to argue for a continuation of this 'special' category, not to support the myth of mankind being the pinnacle of evolution, but rather to establish that we are animals whose 'flashpoints' got so out of control that it required unique anatomical changes and new behavioural techniques to deal with the intractable problems they faced.

This is the reverse of the conventionally accepted position, for when Hobbes referred to the life of natural man as "solitary, poor, nasty, brutish and short," he was suggesting that early man behaved like an animal – "man to man is an arrant wolfe," he said, yet, as we have seen, this analysis cannot be sustained. Hobbes had failed to see he had equated the interspecific act of animal predation with intraspecific human behaviour. My contention is that mankind does 'predate' his own species, not in the technical sense of hunting to kill and eat, (although on rare occasions he does do that) but in the sense of catching and holding others to murder, slaughter, torture, rape, capture, enslave and coerce, that is to inflict a loss of autonomy. It should be noted here that when the victim suffers a protracted loss of autonomy the experience and aftermath can be more terrifying than a swift death resulting from predation.

We do know that throughout the last 10,000 years of human history (and I will later argue, almost certainly well before this date) this fearful behaviour has been a common experience for the hominin. And developing and maintaining defences to these assaults (as well as refining their techniques of perpetrating assaults on others, especially outsiders) has been a central preoccupation of societies trying to establish a civilized social order and it is this genuine fear, I suggest, that prompts the Hobbesian concern for law and order. As Isaiah Berlin says (page 19, *Four Essays on Liberty*), "Whatever his crudities and errors, on the central issue Hobbes, not Locke, turned out to be right: men sought neither happiness nor liberty nor justice, but, above and before all, security." So, if *Homo sapiens* are not enacting inherited patterns of animal behaviour then when, and why, did they adopt these intraspecific assaults that create such difficulties? As Hobbes asked, "Whence the will of mischieving each other?"

Rousseau, on the other hand, believed that, at heart, man was naturally, or innately, good and that he got into these relationship difficulties because of the artificial inequalities brought on because of civilization – inequalities which the social contract was designed to eradicate. However, there is an area of confusion here, for as Maurice Cranston says (page 9, *Rousseau Selections*):

> "...two quite distinct types [of social contract] are in fact depicted in Rousseau's writings: the first is the social contract that 'must have happened' generally at an early stage of evolution; the second is one that would need to take place if people are to live together in freedom."

If there are two distinct contracts, then the history of mankind must fall into several separate divisions:

1. The protohominins who lived as animals.
2. Early man who lived before the first contract, who experienced social difficulties and 'agreed' to become more 'civilized' by adopting the first contracts.
3. Civilized man, who today is still wrestling with social and political contracts in the hope that they will enable a genuinely free mankind, still to emerge.

But these divisions leave no place for natural man, who supposedly lived in total harmony with nature and without social disharmony. This natural man cannot be the man Rousseau called natural or savage (nor is it early or primitive man), for although this man was more in harmony with nature than civilized man, he was not in total harmony, for he experienced the insecurities that the original contract sought to remedy – insecurities that animals who live in harmony with nature do not experience (hence they have no need for any contract). This means that the protohominins (being animals) required no contracts but for some undisclosed reason, by the time they became hominins, they did.

There is another problem here, for while many, including Burke, were convinced that "nature could be identified with the primitive, the chaotic, the sublimely terrible over which civilization created a decent, if frail, veneer", (page 52, *Blackwell Encyclopaedia of Political Thought*) and "that man had passed from the contending disorder of the 'state of nature' to the peace and cooperation of the civil society", Rousseau was not – he thought that "natural man lived in harmony with nature, solitary but free, safe in the hands of nature", "I see him satisfying his hunger under an oak, quenching his thirst at the first stream, finding his bed under the same tree which provides his meal; and, behold, his needs are furnished" (Page 81, *A Discourse on Inequality*, Penguin).

Yet, despite these alleged tranquil origins, mankind was still somehow unable to maintain a social order without a social contract. Rousseau gave two reasons for this, first in the *Discourse on the Origin of Inequality*; he suggested that food resource deprivation, due to overpopulation, caused people to gather in groups to hunt meat and this cooperation showed up the inequalities, which led to social breakdown. Interestingly, this theory of food resource deprivation still holds sway among anthropologists and evolutionary biologists; in fact, it was not until the late 1980s that it was seriously challenged. However, Rousseau had abandoned this theory six years later in his essay on the *Origin of Language* where he suggested that primitive people organized themselves into societies because of natural disasters, such as floods, earthquakes and volcanoes, or because of the harsh climatic conditions of northern latitudes. However, these explanations fail on the simple challenge of why all the other animal species had not responded in a similar way to the hominins (or why did the hominins respond in a different way to all other animals?)

Rousseau understood that natural man must have existed far back in time, "well before the flood", he wrote, and he realized that we would have to wait for more information from the anthropologists before much could be said about our original ancestor. Unfortunately, we still have limited knowledge of the protohominin primate society before the ape/hominin transition, or the hominin societies from the transition up to the commencement of civilization, but the behaviour of extant primates and clues from the fossil record do provide some important information.

In preceding Darwin, Rousseau understandably failed to realize that the hominins had evolved from the primates and that a change as significant as that of 'animal to man' is a major transformation that in so many respects puts man outside of nature, and which is likely to have needed a very unusual selection pressure. For up until the hominins all other animals had adapted over millions of years of evolution to every change of climate, food supply and population density that nature could throw at them. Many of these changes would have created extremely severe pressures yet survival never necessitated the establishment of an enlarged brain nor a self-imposed social contract.

The author of Genesis had, much earlier, also proposed that mankind started out in harmony with God (nature), and she/he suggested that in man's distant past, a major break with nature had occurred, but the break was between natural man and fallen man – natural man living in Eden and fallen man being banished to the 'outside' world. While theoretically there could have been the additional division in man's evolution from natural man to fallen man it now seems much more logical that the break with the primates should also be understood as the 'Fall'. So far, so good, but the religions slipped in a strange confusion here: Jews, Christians and Muslims all believe in an intimate connection between God and nature, but perversely find the close relatedness between animals and humans hard to accept, but as we learn more about our genetic relationship to the animals a more inclusive approach must be adopted. Especially when it is written that when Adam and Eve lived in paradise, they did so without self-conscious reflection, without any guilt about nudity and without toil, because adequate nourishment, shelter and security were freely provided by God (nature). It seems to me that this is a perfect description of the relationship that animals, across all species (apart from hominins), have with their environment, for how else, say, do gorillas in the highlands of the Virunga Mountains relate to each other and exist within their habitat?

I have deliberately mentioned Genesis, noting firstly that it is a literary creation, but secondly that it is also important to try to understand what an early author in this field was trying to express, and thirdly, to draw attention to the perspicacity of this early attempt to understand man's evolutionary history and the remarkable description of what 'living in harmony with nature' entails.

There is also another religious link with this enquiry, and that is with another fictional construct – 'original sin'. Rousseau, believed in the "innate goodness of man", and was uncomfortable with the idea of original sin; he thought that by setting up the prohibitions and demands for obedience, God, rather than mankind, was responsible for the subsequent disobedience. So, he rejected the idea believing instead that the origin of evil was to be found in the restraints imposed by civilization, but, as we have seen with the social contract, this cannot explain the origin of the difficulties that beset early man because these problems preceded civilization and eventually gave rise to its emergence.

St Augustine, on the other hand, championed the belief in original sin suggesting that it was passed from generation to generation upon the seed of the male. Although the Church undoubtedly exploited this doctrine for its own brand of sexual and social control, Augustine was making a telling point for, in linking loss of harmony within nature with the inheritance of the sinful condition, his theory did at least attempt to explain how it was that all of us (but not the animals) fell victim to the universal human neurosis (or sin), which, I suggest, cannot be fully understood until we know why the hominins left the harmony of nature.

Finally, while on this religious theme, we can note that there is a connection between the 'Fall from Paradise' and encephalization, for as Rousseau said, "The first man having received the light of reason and precepts at once from God was not himself in the state of nature."

In conclusion, we can now see that the break between animals and man created an extra division in mankind's evolutionary history that Hobbes and Rousseau did not foresee, so that today's philosophers should have no difficulty in agreeing that the series runs as follows:

> (1) animal primate to (2) early man (who embraces both pre-Stone-Age man and Stone-Age man (3) civilized man and finally to (4) the theoretically possible (much sought after) free, secure, fully autonomous, enlightened man who will emerge in the future (perhaps in Giorgio Agamben's 'community to come').

This series has the advantage that it is the same as that adopted by the biologist studying human evolution (excepting the fact that division four is only a dream, a conjecture, which falls, at present, outside the interests of the scientists.) Note that, in this series, there is no natural man living in harmony with nature, there are the protohominin, *animals,* that lived in harmony with nature, and there are early hominins who, due to unique selection pressure(s), found themselves driven out of harmony with nature to become the hominins and eventually humankind. Thus, we became the only species that require enforceable laws and an elaborate social contract to survive and maintain sufficiently safe intraspecific relationships to enable the females to rear offspring.

The change from alleged freedom to chains at the transition from primitive man to civilized man (the change that Rousseau freely admitted he could not explain – "I do not know", he said) came about only in the sense of there being more chains. These early hominins were not autonomous animals; they were beset with problems and they accepted the chains of civilization hoping to improve (rather than resolve) their social tensions. At no stage in this tacit acceptance of contracts were they in possession of the full facts and origins of their predicament and that is still true today.

The important and tragic change, from freedom to loss of freedom (inability to defend autonomy) had occurred, I suggest, much earlier at the ape/hominin transition, and it is here that we must direct our attention.

Endnote: Prologue

Endnote One:

In a (January 2007) reclassification of primates, (*Nature* 444, p.680) the meaning of the word Hominid was changed to include all the African apes plus humans and a new term 'Hominine' was introduced for the smaller group that comprises of just chimpanzees (*Pan*), humans (*Homo*) and their ancestors.

This has created some difficulties, for while this grouping accurately reflects the genetic similarities between the two species, it does not separate humans and chimpanzees for the purposes of comparative morphology. So, a new word 'hominin' has emerged to be used in a more general sense to cover the group who diverged from the chimpanzees, that is the humans (*Homo sapiens*) and their direct and near-direct bipedal ancestors (*Homo erectus, Homo neanderthalensis, Homo heidelbergensis, Homo ergaster* etc.) plus all of the *Australopithecines* and other ancient forms such as *Paranthropus* and *Ardipithecus*, deliberately excluding the extant and extinct chimpanzees (*Pan*). I shall use the new word 'hominin' when necessary but also resort to the more general terms 'humans and protohumans' whenever possible. (See also, page 17, *Mothers and Others* by Sarah Blaffer Hrdy)

CHAPTER ONE
THE DAYDREAM

On a cold, wet February afternoon in 1973, having finished my farming chores, I shut the office door, sat down, and began to explore an idea that, for many months, had been attracting my attention.

In those days psychoanalysts were suggesting that our individual neuroses were attributable to a lack of 'good enough parenting'. If this was true, the question that then had to be asked was why had improvements in parenting been so difficult to achieve? Critics pointed out that our parents, as children, also suffered from poor parenting, similarly their parents and so on, down successive generations, 'good enough' parents are unlikely to be produced by parents who themselves were 'not good enough'. More than that, these deficiencies were so deeply ingrained in our history that they had become an integral part of human nature; we simply had to accept that we all have a less-than-perfect childhood, get on with it and try to make the most of our lives.

Yet, for many years I had observed the farm animals, as well as the wild animals that lived in the fields around my house, and I had come to see that animals had a far greater capacity for contentment and equanimity than humans. They also seemed to be 'good enough parents' and they did not exhibit anything remotely resembling the restlessness of the human neurosis. If true, these observations opened the possibility that at one period in evolutionary history we were relatively contented animals, and at a later period we had become troubled hominins whose problems and traumas disrupted the animal pattern of 'good enough' parenting to such an extent that it was passed on from one generation to the next.

What could have happened to create such a situation?

At that time, philosophy, theology, psychology, psychoanalysis and evolutionary biology were unable to produce anything remotely resembling a satisfactory science of human nature. As a thought experiment I set out on a deliberate regression, an imaginary journey to the ape–hominin transition, hoping to witness the event(s) that gave rise to human evolution. (In much the same way as we have a rough picture in our minds of the explanatory moment of amphibian evolution, of fish being stranded in muddy pools, scrambling out onto dry land on their bony fins and gasping as they attempted to gain oxygen from the air rather than the water.) Threading my way down the generations I arrived at a point in time approximately 4–6 million years ago at the transition between the last generation of my protohominin ancestors that were still animals (apes) and the first generation of my ancestors that had changed sufficiently to be described as hominins, who eventually would evolve to become *Homo sapiens*.

I imagined an idyllic African forest clearing; beams of sunlight shone through the trees, a group of protohominins were at rest, it was a scene like the chimpanzee family groups that can been seen in all of Jane Goodall's books and films. First, I took the role of a new-born infant clinging to the hair on my mother's body, she was warm and reassuring and I suckled at will. I 'lived' this role for a couple of hours and apart from a few uneasy moments as we tolerated the customary inspections by other females in the group, nothing untoward occurred – maternal care at this stage seemed exemplary.

Next, I moved forward in time to become a four-month-old

infant, my mother was lounging on the trunk of a fallen tree and I was snuggling in her lap. We had a wide area of personal space and there were no immediate dangers. I looked down and saw a small stick on the ground; I thought I could easily scramble down to investigate. I was curious, it was a genuine, innocent, autonomous motivation, well within my powers to accomplish and it presented no threat to me, my mother, or to anyone else. I moved to climb down from my mother's lap and as I flexed my muscles she sensed my intentions, grasped me with her hands and held me to her lap. As I struggled to be free she gripped more tightly, I knew there was no external danger, she knew there was no external danger, it was, on her part, a deliberate and perverse demonstration of restraint and coercive power.

Back in the office, a shiver went up my spine, instantly I came out of the reverie, I knew that I had never seen an animal, of this maturity, held by a parent and restrained in this way, indeed animals with hooves or pads cannot hold in this way. Troubling questions started to form, and I returned to the forest scene…

If I could jump down and pick up the stick and I wanted to jump down and pick up the stick – who then was being held? In some essential sense it was not me, my unitary self had been divided, for I could, and should, by now be on the ground playing with the stick, yet here I was still sitting on my mother's lap. And how could a mother become so divorced from her programme of maternal care that she would indulge in a perverse demonstration of power over her own offspring?

This was an epiphany for me; I was convinced that this event needed further exploration.

Over time the real-life implications of the use of hands to grasp and hold gradually opened into a series of fertile concepts, to the extent that later, in a second daydream, I was able, to regress again and 'see' the ape/hominin transition from the adult female perspective and to start to understand the difficulties she would have experienced living in a world in which she was subject to severe restraint.

Before we take that imaginary journey, it needs to be set within the context of animal sexuality. The female animal is able, without any assistance, to protect herself from the males of the species. She has a short oestrous period in which she is hormonally prepared to mate and a longer anoestrous period in which she is genuinely, absolutely, safe from the attentions of all the males of her species. It is her own hormones that determine when she can, and cannot, mate (note it is not her conscious mind, nor is it the male sexual drive that makes those decisions). This system is the basic pattern within nature and it has protected all female animals over hundreds of millions of years from the undue attentions of the male.

If we now go back about 4–7 million years to an area of Africa (or whichever continent the arboreal/terrestrial transition first took place) that was originally densely forested, we find that after a period of prolonged climate change many of the trees have died and those few that survived have far fewer leaves and fruit. Therefore, crowding, stress and hunger would have driven some of the protohominins to move out onto the savannah to try to live off the sparse shrubs, plants, roots and long grasses.

Now imagine that you are an adult female protohominin trying to survive in these harsh new conditions, you are hungry, you leave the area around the few failing trees that give you some protection and venture out, with others, onto the open plains. Two days ago, you were in oestrous and willingly mated with the males. Today you are certainly off heat (anoestrous) and in need of finding a nourishing supply of food, so you confidently move to an area that you know has some low-growing berries. All seems well, the berries are ripe, and you eat contentedly but one of the males you mated with two days ago is behaving strangely. Surely, he cannot think you are still in oestrous, can he? No, you made it perfectly clear through your actions, your body posture and your lack of pheromones that you are certainly anoestrous – 'off heat'. Nevertheless, he keeps encroaching, he has an undue interest and he is not respecting your body space.

It does not make any sense. Instinctively, you look for a tree, you know you are always safe in a tree but there are none. You move on, he follows, you run, he runs. You stop to confront him, you snarl bark and squeal, it seems to make no difference, he is intent on bearing down on you. Why is this happening? You have never experienced anything like this and you don't know what to do. If only there was a tree or a cliff edge, then you could escape.

He catches you with his hands, grips you, holds you, you wrestle and try to bite him, but he is stronger and heavier, he holds your arms, his grip tightens, he jumps on your back, he presses you to the ground and penetrates you. You do not know this, but you are a member of a species that has the most vulnerable females in evolutionary history and you have been coercively mated, even though you are certainly anoestrous. You wriggle, fight, rear up, squeal, but to no avail, you try to bite but he has positioned himself to the rear and he can bite harder than you and his teeth are now close to your neck. Other males see and hear the confrontation and they move closer to watch your predicament. Eventually he tires and releases his grip; you take your chance and run away.

The three watching males now join in and chase after you; you get just twenty metres before the largest male, with his outstretched hand with its opposable thumb, grips and catches hold of you and you are brought to a juddering halt. For the first time you realize a terrible truth, out here on the plains you are defenceless, utterly defenceless, the males can catch, hold and motor-control you whenever they choose.

Your immediate concern, quite rightly, is for your dreadful plight but there is also a larger issue here: the predicament that has overwhelmed you will also go on to overwhelm all mothers, daughters and their daughters too. The basic unpalatable truth is that, unlike the vast majority of animal females, hominin females cannot, single-handedly, defend their own autonomy (see Coercive mating, Chapter 2:6).

This is the beginning of a monumental arms race between the hominin male and female. It has set in motion a confrontation that has spawned many defences and counter defences, but it has remained unresolved for more than five million years and we are still trying to come to terms with it today. The unfolding of this arms race is the story of human evolution. This book, *Hands, the Achilles' Heel*, attempts to outline that process.

CHAPTER TWO
THE BASIC CONCEPTS

It is generally agreed that our hominin roots go back about 5–7 million years to the time we diverged from a relative of the chimpanzee. Thus, we originate from the great apes with their brachiating abilities that were perfectly adapted to an arboreal lifestyle, where the design of their shoulders, arms and hands gave them the necessary agility to swing through the forest canopy. However, sometime following periods of climate change 3–7 million years ago, [1] when the forested areas were significantly reduced, some of the apes were forced to leave the safety of their arboreal or, semi-arboreal, lifestyle for a new predominantly terrestrial existence.

This arboreal/terrestrial transition had the unexpected consequence of freeing the hands of the apes from the constraint of holding onto the branches in the trees. Most theorists including J. Napier (1980), R. Tallis, (2003) and C. McGinn (2015) see this as providential, for it allowed the protohominin/hominin hand to be used for other purposes such as carrying food and making tools and weapons, which, in turn, led to the enlarged brain, culture, religion, farming and eventually to writing and the accumulation of modern-day knowledge. And they do not hesitate to claim that these remarkable achievements selected a superior intelligence that has projected humankind to the 'pinnacle of evolution'. However, this view of human exceptionalism overlooks one crucial factor that, once fully understood, will lead to a re-evaluation of the way these developments will be assessed in the future.

That crucial factor is motor-control.

2:1 Hands and Motor-Control

When our ancestors came down from the trees to become ground-living protohominins, they inadvertently became exposed to an entirely new danger. I suggest, the selection pressures arising from this danger have had such far-reaching consequences, even to the extent of modifying our anatomy and physiology, that it is difficult to overstate its effect. Yet remarkably, this danger has remained undisclosed even though we experience those effects and consciously plan defences to them every day of our lives.

This danger is the loss of autonomy, the coercion and restraint that arises from the use of hands to motor-control others. It came into existence as the protohominins were forced to adopt a wholly terrestrial lifestyle and the females lost their place of safety in the treetops.

Immediately the protohominin hand is freed from the constraints of holding onto the branches in the trees it can be used for other purposes and while we have concentrated on the hand's ability to hold and grasp, which facilitated the skills of toolmaking and tool use, we have overlooked the fact they can, and surely will, also be used to grip and hold the limbs, torso and hair of other protohominins.

This vulnerability comes about without access to trees, which are a place of safety, and even in a semi-arboreal lifestyle any potential attack could easily be avoided by the simple act of running up the nearest tree and going out on a small branch capable of holding the weight of only one, where the attacker dares not follow for fear of falling. Once this arboreal safety is no longer available then, I suggest, the unsolicited use of hands becomes extremely threatening and it opens an unprecedented ability to restrain and coerce that has dislocated the long-established patterns of safety that previously regulated the coexistence of individual animals of the same species.

The moment the controlee is held against her/his will and her/his movements are directed by the controller then she/he experiences heteronomous motor-control. This restraint must axiomatically be experienced as a major threat because it results in the controlee suffering a debilitating loss of autonomy. Put another way, motor-control means that the controlee must have been taken past her/his point of submission and is unable to implement any avoidance strategy. Unsolicited motor-control is always a trespass imposed by others.

In this context, 'motor-control' means having the power to control and trammel the movements of another individual by holding, hitting, grasping or pulling. And throughout this text, the brief term 'motor-control' will mean 'unsolicited, intraspecific, [(2)] inter-individual, heteronomous motor-control'.

Motor-control acts as a powerful selection pressure because it overwhelms the individual's own autonomous control of movement, and it therefore creates a loss of autonomy in the controlee, by overriding the escape and avoidance responses that otherwise she/he would have used to evade any danger. Its most disruptive manifestation, dating back to the protohominin/hominin transition, is coercive mating (rape).

The hand [(3)] becomes a weapon of great power when it is used to grasp, hold, manipulate and control another individual, but the actual use of the hands, in the act of motor-control, does not cause serious physical injury (unlike the jaws of a predator). Hands are relatively gentle compared to teeth, claws and horns [(4)] and so the victim will usually survive the assault (unless the attacker is bent on maiming or killing). But survival means that she/he will be left having to deal with the mental trauma that arises when autonomy is lost because she/he has been held, coerced and injured, and left having to find adequate defences to avoid further attacks in the future.

Thus, the ability to use hands for intraspecific motor-control should be understood as the Achilles' heel of the hominin species since it exposes a virtually indefensible weakness in the individual's defence of autonomy – a weakness to which we are all vulnerable.

To fully comprehend what occurs within the individual when he or she is motor-controlled (grasped, held and manipulated) it is necessary to understand how autonomy functions and what happens when it is lost.

2:2 Autonomy

It is easy to say that unsolicited motor-control produces an instant loss of autonomy in the one who is controlled, but that statement masks the fact that we have a totally incorrect understanding of the term autonomy and little understanding of the fact that much of human governance is under heteronomous (rather than autonomous) control. In this reappraisal, I will set out what I mean by the terms to 'have autonomy' and to 'lose autonomy'.

First, it needs to be made clear that the 'body autonomy' that is violated by motor-control is an autonomy that lies at a much deeper physiological level than the autonomy we call 'Kantian moral autonomy'.

Kantian moral autonomy is a property of the will of a rational being (which has a large heteronomous content) that is thwarted when conscious top-down plans are denied. Whereas, the maintenance of basic 'body autonomy' is the principle by which all animals determine and grant priority to the behavioural programme (motivational system) that best protects its autonomy at any given moment. Basic body autonomy is lost when the individual is denied the ability to express its own priority 'bottom-up' responses selected by its own safety/danger scales. (See Chapter 6.)

This lack of clarity about the function, extent and importance of true autonomous governance has bedevilled philosophy and the natural sciences from the beginning. We will make no progress in understanding human nature until autonomy has been thoroughly reappraised – hence the subtitle of this book, *Towards an Understanding of Autonomy, Heteronomy and Human Freedom.*

Autonomy is said to mean 'giving oneself one's own laws' but that is not an accurate definition, for all individuals are constrained by the physiology and behavioural patterns that they inherit from their own species. The individual does not create, construct, or give itself these laws (behavioural patterns), it simply abides by the bottom-up programmes that are given to it at birth. Autonomy is the ability of the individual to freely enact the response programme that is chosen by its own bottom-up safety/danger scales. The individual must continuously make bottom-up priority decisions to keep self-alive and these can only be enacted by a fully autonomous individual. So, abiding with and respecting one's own laws means abiding by the behavioural responses of the deep body organization with which one is born. It does not mean enacting the top-down rational decisions that an individual human makes, because many (most) of these decisions have a large heteronomous content. Immanuel Kant was aware of the heteronomous content of many decisions, but he was unable to grasp the deep biological basis of autonomy, which allows a self-sufficient individual to enact the bottom-up priority decisions of its own safety/danger scales. Trying to find pure autonomous activity in the top-down rational will (that is born out of heteronomous controls) is a misguided and impossible task.

2:3 Priorities within the motivational system – safety/ danger scales:

The maintenance of autonomy needs to be understood as one of nature's most fundamental principles. All animals organize (govern) their own behaviour and in so doing they continuously have to make priority decisions that determine their next movements.

Animals are born with a species priority system (safety/danger scales) that autonomically selects the most appropriate behaviour for them to enact what protects them from the most urgent threat that they face at any given time. Once that threat is dealt with, say, fleeing the predator, then another threat takes priority, say, hunger, and once that is satisfied, resting or sleeping to conserve energy may take its place, and so on throughout each day of their lives (deviation from this priority rule quickly leads to death; clearly if you are a prey animal it is extremely dangerous to continue grazing while the lion approaches).

Priorities have been determined by evolution, for although animals (including mankind) have motivational systems at their disposal, which deal with all the threats that the body is likely to meet (such as thermoregulation, energy inputs, water balance, waste disposal, exhaustion, maternal care, social cohesion and avoidance of external dangers) they have only one set of limbs and/or mouth with which to execute them, so generally only one output (in addition to breathing) can be attended to and carried out at any one time.

The result is that an autonomic (bottom-up) choice must be made, based on principles of safety/danger [5] selected over years of evolution, as to which motivational system is granted priority. It is crucial for safety and survival that these choices of priority can be freely expressed as and when they are required. This internal, subconscious, bottom-up governance is how basic body autonomy manifests itself.

To be able to prioritise decisions in this way, a large volume of information must undergo a complex internal processing, referencing and interchange. Such things as, the levels of nutrients, hydration, hormones, waste products and energy reserves, as well as health, status, age, size and gender, together with information about present external circumstances, memories and future requirements, must be speedily monitored, registered, and cross-referenced so that an appropriate (tested by evolution) response can be set in motion. Thus, by definition, this internal, bottom-up information is available only to the brain/body of the individual concerned. It is impossible for any other individual to have access to the depth and complexity of this information, so autonomous governance is the only basis upon which safe behavioural decisions can be made.

To illustrate how this internal knowledge can only be available to the individual's internal system of control (and to none other) consider the behaviour of the toad-eating snake *Rabdophis tigrinus* (*New Scientist* 03.02.07). Toads have a poisonous skin that deters most predators, but the toad-eating snake can eat them without harm. More than this, the snake then uses the toad's poison to create its own venom. If the snake has eaten many poisonous toads, it will have a store of venom which enables it to attack aggressively. If it lives in an area where there are no toads to eat it will lack venom, so its behaviour will become less aggressive as it will not be able to poison either its prey or its predators. In transitional areas it depends on how many toads it has eaten whether its defence is active or passive. When it decides to attack (or not), no other snake, no other creature, no other system has access to the necessary internal information on the levels of stored poison in its body. Hence to ensure survival it is imperative that the individual relies on its own autonomous decision-making process. If heteronomous control was somehow possible in this situation, how could it be privy to the information necessary for efficient internal governance?

Similarly, with humans, a parent, for example, may put out a large meal and then insist that their two-year-old child eat the whole portion but without the necessary knowledge of the child's physiological state at that moment in time; the parent simply does not have the necessary information to make a correct decision.

Now we can see that there is a clear difference between the internal governance of the bottom-up autonomous animal outlined above and the common human experience of being subject to top-down heteronomous external controls – "Come here," "Sit down," "I told you not to do that," "Eat that," or "Do your homework." Having to enact the commands of others in this way is a major difference between mankind and the animal, and the implications will be discussed throughout this book.

Unsolicited motor-control jeopardizes the internal organization of the controlee because it overrides the basic bodily autonomy of the individual by forcibly redirecting the control of behaviour from the original internal (autonomous) source to a new imposed external (heteronomous) source.

There is no possibility of heteronomous controls taking total command, for the reasons outlined above, so when they are imposed, a dual control system is created that has, somehow, to be operated by the controlee and this always has a considerable cost. It is their vulnerability to unsolicited motor-control that makes the hominins the only species that has to operate this dual system of governance and I suggest that this is the selection pressure that has driven the need for the extra capacity of an enlarged brain.

2:4 Loss of autonomy:

The hominins inherited physiological structures and processes like those that maintain basic bodily autonomy [(5)] in all other animals because they shared most of the ways in which that autonomy can be lost. For example, both a hominin and an animal must avoid starvation, heat exhaustion and dehydration. In addition, they must avoid being trapped by a rockfall, a landslide, in a thicket, in quicksand, or swept away in fast-moving water, and both must avoid being caught in the jaws of a predator. Being held in these ways will usually result in death, so the selection pressure favours those who rigorously avoid these situations.

These threats to autonomy are relatively common and, harsh as it is, death does resolve the victim's existential difficulties. If the victim survives then it will have learnt to show more caution in the future but basically it continues to rely on the species' autonomic defence programmes because, over evolutionary time, they had proved to be the most successful.

However, the protohominins experienced something that was uniquely different: when they were held with 'gentle' hands they physically survived the act of being held to find that their own autonomous control had been taken over by the controller, who then went on to coercively mate and/or direct the controlee's behaviour upon the threat of punishment.

Motor-control causes irreparable damage to the organization of autonomous governance and this loss of liberty must be registered by the body's defensive system as the most dangerous non-fatal situation that can be experienced, and consequently it is reacted to with extreme distress. The difficulties of this situation are compounded by the fact that when autonomy is threatened by individuals from within one's own species, it is impossible to totally avoid the threat because proximity and contact (at least some of the time) are a necessary part of the essential biological functioning of all mammalian species.

Mating and maternal care are examples of two vital biological programmes that mammals cannot avoid or significantly alter that, by definition, involve close contact. If the dominant partner in these relationships can motor-control, then an approach/

avoidance conflict is created in the controlee that produces a permanent ambivalence, which can generate a dynamic powerful enough to drive hominin evolution (see Chapter 3).

In considering these problems it should be remembered that over the years of hominin evolution our species has been selected to tolerate external restraint and heteronomous coercion. That is, we have become domesticated, or put another way, only those who were passive enough to endure the coercion and compliance of motor-control were able to survive and go on to have progeny. Thus, being forced to accept heteronomous controls meant that a domestication of the brain had to be selected that modified the animal's imperative to always defend autonomy. This once-essential reaction, (part of our animal nature and natural wildness) has been compromised and in this essential way we became different to all the other animals. (Originally the protohominins were untrammelled wild animals and, like any wild animal today, they would have been intolerant of control and coercion. Indeed, you cannot get close to most wild animals because they do not allow even their 'flight distance' to be intruded upon.)

Much of our acceptance of many facets of human behaviour, such as discipline, obedience, control, coercion, incarceration, compliance and the ability to follow instructions, must be seen in the light of the fact that we have been selected to become more tolerant of motor-control and coercion. So, to achieve a deeper understanding of what it means to lose autonomy and accept heteronomous control, we must be able to think through our own domestication to see how our animal ancestors experienced the world and therefore see how distorted, strange or unusual our present-day 'normal' human behaviour has become.

We can get an insight into this situation by observing domesticated animals that have been selected for an ability to tolerate or endure the most unnatural conditions. Dogs may be placed on the lead and taken to obedience classes or shut in kennels, horses are 'broken', harnessed and shod, subjected to the whip and the bridle, hitched to the cart or plough and shut in stables, these are all experiences that their wild cousins would find intolerable. Without the human use of hands and the aid of restraints such as ropes, leads, halters and fences, the animal would not succumb to this level of interference and without the selective breeding of domestication they would go berserk and die.

Back to the problem faced by the protohominin female out on the savannah and assaulted by the male: if she survived, she was left in a state where her instinctual, bottom-up programmes for defence had been overpowered. Without assistance she would always have to comply with the top-down heteronomous directions of her controller and succumb to his penetrations even though she was anoestrous.

Later she would try to improvise her own top-down plans and adopt behaviour that, while complying with the heteronomous controls, would also try to mitigate her plight. In so doing she found herself (in evolutionary terms) in an extremely rare (if not unique) situation which has driven hominin evolution ever since.

2:5 The oestrous/anoestrous cycle:

Mammalian sexual reproduction has followed a broadly similar pattern across a wide range of species for millions of years. The male must deposit his spermatozoa inside the female's body where it can fertilize her ovum. There are some inherent difficulties with this system because male and female interests are not identical, and this produces conflicts and compromises throughout the process of mating and conception.

Under the hormonal changes of oestrus, the female offers herself for insemination, she stands on 'heat', allowing the male a short period of access for the intromission of his penis. It is in this limited period that the female usually sheds her ovum and is at her most fertile, thus the male must be prepared to mate as soon as the female comes into oestrus. Oestrus is determined by the female's own hormones that make her ready, willing and prepared to mate and its duration varies between species. The mammalian oestrous period is relatively short, often two days, sometimes up to seven days and occasionally over twenty days in the case of bonobo chimpanzees. It can be a stressful time for the female who endures male buffeting and mating activity throughout the whole period, often from more than one male.

The key point here is that as soon as the female's hormones subside (which takes six to twelve hours) the oestrous period comes to an end, the female then becomes anoestrus, she is no longer sexually receptive and her fertility declines. If the mating results in conception, anoestrus will last (depending upon the species) for months or years, throughout pregnancy and lactation, after which she will return to her oestrous cycle and breed again. In the case of the chimpanzee (our nearest relative) anoestrus will last for over four years and for the orangutan it is six years. If conception fails, the female will return to the oestrous cycle every three to four weeks (actual time depending upon the species) until she becomes pregnant or barren. In the six to twelve hours that the female takes to 'come on' or 'go off' heat, her hormonal status is uncertain and as the female comes under pressure from the male she will show her reluctance with squeals and agitated movements, but this is a very short-lived period. It may be mistaken for male aggression, but it is an inevitable and natural consequence of making such a major change in behaviour in a very short period.

Oestrus determines the duration of the period within which sexual activity can take place, for once oestrus ends, the female becomes anoestrus. Once the female's oestrous hormones subside the male is no longer able to mate and in most species, it has not been within his power to change that situation. Most females know that once they are anoestrus they are guaranteed safety, for either the male does not try to mate, lacking the stimulus of the signs of her oestrus, or he knows that he cannot mate because the female will move away as he tries to mount, and he will be thwarted. In fact, the reason the male animal can inherit his instant one hundred per cent mating response is that the female is able to regulate his behaviour by restricting access to her body to the relatively short time of her oestrous period, for once she goes off heat and becomes anoestrus the male cannot mate even if he was keen to do so. The female is always under pressure from males seeking to gain any advantage over their fellow males and without the safety barrier of anoestrus, which is established as the limited period of her oestrous ends, she would be overwhelmed. The practice of having a short period of oestrus and a guaranteed longer period of anoestrus has stood the test of millions of years of mammalian reproduction and throughout this time it has been remarkably effective in protecting the female from the male's sexual drive.

The hormonal changes of the short oestrous period allow the female to become tolerant of quite rough male behaviour that she would normally avoid. When male behaviour becomes too rough she will (if physically able) override her own mating hormones and ameliorate her discomfort by moving forward to dislodging the male off her back and he will not be able to mate until she stands again (allowing him to mount). The degree of forcefulness that the male exerts varies between species and it is a major factor in the extent of difficulties that the females may experience while she is in oestrus. The harshest behaviour of this type is reached when the female is physically held so that she cannot escape from the male and he is able to position her for insemination (even if she is not ovulating or fertile). In coercive mating the female is compromised; she is overwhelmed, and her autonomy is violated. The female is not an inert substance, she is an autonomous living being that must adjust to and accommodate intromissions into her body so that she can breed, but there are dangers and physical limits to this procedure that she needs to guard against.

The ability of the female to establish anoestrus is crucial to her autonomy and it is a central feature of this discussion. Observation of the transition from oestrus to anoestrus will highlight the difficulty (or not) that she has in this area and show to what degree and for how long she must compromise her behaviour to accommodate the male and at what point this compromise becomes intolerable. All females will evolve countermeasures to minimize the effects of coercion. Any change in behaviour or anatomy that affords protection to the female, that maintains or improves her safety and increases the chances of her leaving viable offspring, will be selected.

2:6 Coercive mating:

Coercive mating is central to the hypothesis discussed in this book, so the essential features are examined now in more detail. There are some non-hominin species in which the female has become very vulnerable to the male while she is in oestrus and coercive mating has been documented in a few species including insects, birds, amphibians, mammals and primates. (7) In these cases the male has developed the ability to hold and/or manipulate the female. Some insects such as scorpion flies have grasping appendages on their hind legs, frogs and toads can grip the female tightly with their arms, some ducks can hold the female under water, horses can bite and corral the females, primates have prehensile hands and the orangutan has prehensile hands and feet that facilitate coercion. Much of this coercive activity takes place in or around the time of the female's oestrous period and however harsh the experience is for the female it is time-limited, for when she regains anoestrus she is safe from male attention. The establishment of the basic oestrous/anoestrous cycle has been a constant feature for millions of years that has kept the female safe from male attention. (Indeed, in most species, if the anoestrous female meets a species male she will not fear for her safety for either he makes no sexual advances, or she is genuinely safe from any advances he may make.)

Hence, there is a gradation of mating difficulties experienced by the females of different species, from the ovines and bovines (with hooves and no canine teeth) in which mating is virtually stress-free, to the orangutans (with their prehensile hands and feet) where the females experience the most forceful mating of any non-human animal. At the extreme end of this gradation are the hominin females who, remarkably, are the only species to have lost their ability to regain anoestrus naturally and this is the starting point of a very serious biological problem. For it means that the human female is no longer safe from male attention and will fall victim to the male any time the male chooses, and she will suffer the serious physical and psychological damage that follows in the wake of coercive mating (rape).

It is only by understanding this transition from animal sexual activity (based on the oestrous/anoestrous cycles where the female is always safe) to the sexual behaviour of humans (where the females can easily become unsafe) that we can see why (and how) we have evolved to become human beings reliant on social and sexual contracts.

It is in the female animal's interest to advertise her oestrus to be able to mate with the fittest males. If oestrus is not advertised, then this fact suggests that she is having difficulty maintaining her autonomy when faced with male attention. For example, orangutan females are known to experience the most aggressive animal coercive matings (Smuts and Smuts, 1993) and they are the only primates (apart from the hominins) that do not exhibit a genital swelling to advertise oestrus. However, they do attract the male with oestrous pheromones and they do cycle between periods of oestrus and anoestrus, but this is often a weak anoestrus because the female can experience being mated while she is pregnant. This is a sure sign the relationship is under stress from male aggression. (Orangutan and bonobo females may acquiesce in some of this behaviour as it is often in her best short-term interest to placate the forceful male and make bonds and alliances with him rather than suffer more severe violence but this capitulation is at the limit of the female's tolerable, functional, autonomous behaviour.)

By comparison, human females are permanently receptive (and ovulate silently), which means they do not exhibit the oestrous/anoestrous cycles in the standard animal way. This odd, untypical behaviour has, I suggest, come about because the hominins, having left the safety of the trees for a terrestrial lifestyle, became vulnerable not only to coercive mating but to being motor-controlled, gripped and held at any time by the hands of their conspecifics. What must be fully thought through are the consequences of this vulnerability, for once the boundaries of the oestrous/anoestrous cycles are violated it is difficult to see how the female can function autonomously.

Although the females of some animal species, such as the orangutan and bonobo chimpanzee, do suffer coercive mating, they have not lost their oestrous cycle or their ability to establish a functional anoestrus.[8] All of which implies that their experience of coercion is not as severe as that experienced by the hominin females. It seems that the arboreal primates gain a considerable degree of protection by being able to escape up into the forest canopy, which means that the coercive attacks on the female are sporadic (rather than constant, widespread and common). When the hominins became terrestrial they lost their best natural defence in the treetops and they were left with weak protection such as hiding in caves or thick undergrowth, wading out into deep water or sitting on cliff edges, but not much more. I suggest that the coercion experienced by the hominin female became constant, widespread and common, and this inevitably led to the loss of her ability to commence and terminate the oestrous and anoestrous phases of her cycle, which in turn necessitated the development of the unique hominin social sexual relationships and contracts discussed in the following chapters.

Thus, mating tensions are apparent in several species but even female orangutans, who have the most severe difficulties of any animal, have clung to their anoestrous period and are undeniably part of the animal kingdom. Whereas, I suggest, the mating difficulties faced by the hominin females were so severe they had to abandon their overt oestrous/anoestrous cycles, which in turn has necessitated such major behavioural adjustments that we now consider ourselves to be different from the animals.

This biological change has been of such importance to the course of human evolution that I will recapitulate by setting out the gradation of difficulty that can be experienced by the females of different species, showing that they fall within one of three divisions:

1. In the clear majority of mammalian species, anoestrus provides a guarantee of safety to the females and the males comply to the dictates of the females' hormones. As the female comes into oestrus she 'stands' so that the male can mate, as her oestrus ends she becomes anoestrus and does not 'stand' and the males accept the fact that they can no longer mate. The females in this group are not seriously threatened or coerced at any stage of their oestrous/anoestrous cycles.

2. In the few species that have 'coercive' males, the females have a more difficult time, for example, mares may be bitten and corralled by stallions and primates may be gripped and held by the male. Although the females in this group may be coerced

when they are in oestrus, they are not overwhelmed, for within a few days they are usually able to establish anoestrus. In a small number of species within this group the female's anoestrus may not be fully functional (as in the case of the orangutan or bonobo) and she may be mated while pregnant, however, her anoestrus is adequately functional, in that she is not overwhelmed, and these species, despite their difficulties, have remained classed as animals. But they clearly illustrate the difficulties that coercion creates for the female. This group can be understood as occupying an intermediate position between those in groups 1 and 3.

3. The final group consists only of the hominins who have crossed a threshold of coercion. The terrestrial hominin female is so vulnerable that she can be coerced irrespective of whether she is in oestrus or anoestrus, and so it became essential for her to change her behaviour, to seek new defences, including social and sexual contracts, to achieve some protection. This is a truly remarkable situation for it was probably the first time in mammalian evolutionary history that the female was left without any ability to defend herself and establish anoestrus (that is, she remained permanently receptive). This is not a sporadic short-term difficulty that only occurs in or around the hominin female's oestrous period, it is a continuous long-term vulnerability of the female to motor-control and coercive mating. In this situation the hominin female's anoestrous cycle is made irrelevant because it is so easily overridden by the males. It is clear from the fact that all other animals have retained their oestrous/anoestrous cycles (where the hominins have not) that the terrestrial hominin females were uniquely vulnerable to coercive mating out on the treeless savannah.

Once the hominin male realizes he can violate the female in this way it is likely he will do so time and time again. This becomes a regular lifelong vulnerability from which the female cannot find sanctuary without assistance. This vulnerability to repeated rape (and ensuing injuries) becomes a powerful pressure that selects any behaviour that reduces the violation. Even if that means the female needs to form alliances, and endure restrictions, that would have been intolerable and unnecessary when the species females were able to establish anoestrus. Those who have found ways to avoid the worst excesses of coercive mating, by cooperating with a limited number of males, in exchange for protection from most other males, will be favoured and those that did not make accommodations of this type would be selected against. This is a powerful pressure that acts on the female line and any improvements in safety would quickly be selected.

Oestrus prepares the female for mating, she expects (wants) to be mated and she knows when she has been mated. She also knows when she is anoestrous and if she is coercively mated she knows that she has been restrained, or threatened, to facilitate penetration. Being restrained is a fate (irrespective of any sexual content) that is resisted by all autonomous wild animals for it is debilitating to their autonomous function. Thus restraint, and/or coercive mating, is experienced as a loss of deep-body autonomy. It is not a top-down conscious decision of the rational will to choose to dislike this experience, it is an essential function of the bottom-up safety/danger scales that ensures that it is always avoided. There is a difference between coercion while the female is oestrus and coercion whilst she is anoestrus. Coercion within the oestrous period, when the female is primed to 'stand' for sexual activity, seems to be less damaging than coercion when she is anoestrus because her body is not primed in any way for sexual activity.

If autonomy is threatened, or overridden, then the controlee must adapt in the best way that is available. Hence some primates mate out of cycle and female hominins have become permanently receptive but the fact remains that if the males were unable to mate in the anoestrous phase of the female's cycle (as is the case with most animal species) then these difficult pressures on the female would not occur.

The females that do allow mating when they are anoestrus can be seen to be under pressure and they have learnt that if they present to the male and comply outside of the oestrous period they can reduce the overt physical coercion they are under and they may even be able to gain favours from him later. In these circumstances we see that that the female improves her difficult situation by enacting this strange out-of-cycle behaviour. Restricting mating to the oestrous period has been the norm for the vast majority of female animals over millions of years of evolution and any deviation from this pattern must have been due to the female's inability to maintain her autonomy and establish anoestrus.

2:7 Male sexuality:

The male of the species has a powerful sexual drive that has evolved to continuously monitor the females for signs of oestrus and instantly respond whenever a mating opportunity arises. This instant, urgent response is an inevitable outcome of the dynamic of intra-male competition responding to the female reproductive cycle where the winner, in terms of gene survival, is the first sperm to reach the ovum.

The female knows that when she is in oestrus a male will appear. Not all males respond to mating opportunities one hundred per cent of the time, some will be sick, injured or inhibited, but in each area, most males will respond and be ready to mate one hundred per cent of the time. Sometimes the female may show a preference for a certain male(s), but this makes little difference to the underlying pattern, for the chosen male still must show an instant response.

This powerful male drive gives rise, in a few non-hominin species, to a situation where the females can experience what appears to be coercive mating but generally this is aggressive mating that takes place within, or at the end of, the oestrous period. These cases, while stressful for the female animals involved, do not amount to the widespread vulnerability to coercive mating that the hominin females experience and none have lost their oestrous/anoestrous cycle. The species in which coercive mating has been reported are discussed in Endnote 7.

When the protohominins became ground-living primates, they were still animals subject to standard animal sexual responses. So, when the males, with their hands free from arboreal constraints, found that they could select, catch, hold, molest and gain advantage over the females whenever they chose, the scene was set for a seismic conflict between the male sexual drive and the female's need to protect her autonomy and it created an intense selection pressure on the female line. It is hard to overestimate the consequences of this conflict for it is the underlying story of hominin/human evolution.

2:8 The inherent conflict between female and male sexuality:

T. Clutton-Brock and G. Palmer say in their paper in *Animal Behaviour* 1995:

> "In a wide range of animal species, males coerce females to mate with them, either by physically forcing them to mate, by harassing them until they mate or by punishing persistent

refusal to mate… this paper argues that the possibility of forced copulation can generate arms races between males and females that may have substantial costs to both sexes… it is suggested that sexual harassment commonly represents a 'war of attrition' between the sexes."

This is the present orthodox scientific position, but it does not contain within it an understanding of the use of hands to motor-control. Hence it is limited in its understanding of human sexuality. Even though the females of some species are harassed the reason why most female animals are not seriously threatened by the male is that, in most species, the male animal simply does not have the ability (without the female's hormonal cooperation) to make her 'stand' to mate even while she is in oestrus and certainly not when she is anoestrous. This is because she is always able to establish safety by means of avoidance, by the simple procedure of moving away. All that the quadrupedal female must do to thwart the male is to move two steps forward as he rears up to mate and all the female primate needs to do is move to the end of a branch, or onto a cliff ledge, where it is too dangerous for the male to approach.

When the female is ready to breed (fit enough to ovulate) she will signal her condition (end of anoestrus) with a period of overt oestrus. This shows that she is 'on heat', her body is hormonally prepared, first for the rigours of male attention and then for pregnancy, and she shows this by being willing to 'stand', then, and only then, is the male able to mate with her. As her period of oestrus ends her hormone levels decline until she is once more anoestrous (and probably pregnant) and totally safe from male sexual pursuit. For the female animal who can easily establish anoestrus, oestrus is experienced simply as a relatively short, unproblematic phase of her reproductive cycle.

A remarkable inversion often takes hold at this point of the discussion where the benefits of oestrus are denied, or misconstrued, by humans who see oestrus as being out of control, bestial and in thrall to the blind lust of hormones. Yet if an overt oestrus were, somehow, to be returned to the human female it would signal that she had regained her autonomy and the ability to establish anoestrus and, with it, her ability to easily and naturally defend herself from coercive mating and the heteronomous controls of the male. It would restore a basic freedom and autonomy that today she does not even dare to think should be available to her in the way that it is to all other female animals.

When male animals attempt to mate they are (in part) stimulated by the stationary nature of the female when she 'stands' on heat, signalling that she is available. It is unfortunate that this stationary state is replicated when hominin females (and males) are held and motor-controlled as this leads to an added complication. The stationary aspect of control tends to excite the male into sexual activity (or violence, as displaced sexual activity) and so it compounds the problems that arise when dealing with the non-sexual aspects of motor-control.

As soon as the male primate/protohominin's hands were freed from the constraint of living in the trees, then grasping and holding the female and coercively mating (raping) her became relatively easy, it was (is) simply a question of male strength/power overriding the female's anoestrous cycle. Obviously, there is an ultimate constraint on male-on-female violence for the male's genes cannot survive unless the female is able to breed successfully. The most violently aggressive males would be selected out leaving those who used their hands less aggressively to restrain and control the female. However, although this coercion is apparently more gentle, in the sense that it causes less physical damage, it still causes considerable psychological harm. For every time the female is mated when not in oestrus, she suffers a serious loss of autonomy, even if she has been heteronomously persuaded to participate.

Eventually, cooperative defences (see Chapter 8) developed

social and moral codes to restrain the rampant male. However, these codes of behaviour require constant reinforcement as can be seen when law and order break down and the male senses that he will not be caught or punished, as happens with conquering armies, lawless gangs and lone sex offenders.

Recently it was reported in *The Times* newspaper that a fifteen-year-old girl was abducted from Moldova. She was taken to Italy where she was sold to a pimp who, in the first six days, with others, raped her continuously. Later, she said she was unable to stand or walk for a week. This is in twenty-first century Europe with a sophisticated, civilized legal system and with an active, well-financed police force. In contrast, not a single fit female animal, of any species, throughout millions of years of evolution, has ever suffered such a debilitating ordeal perpetrated by the males of her own species.

This harrowing report is included here to illustrate that if the human female has no assistance, or protection, then she is extremely vulnerable and if attacked, her health, breeding status and survival are severely compromised; usually the victim survives the attack (unlike predation) to experience the trauma for the rest of her life. Rape disturbs the female's well-being to such an extent that it is likely to disrupt the close bonds of motherhood that are essential for the longer-term survival of the family. Rape may result in pregnancy, but it is not conducive to well-being and maternal bonding. Females who are unable to avoid the trauma of rape will be weakened, whereas those who are able to avoid it will be favoured by selection.

2:9 Permanent receptivity:

I suggest that as soon the hominin females were denied the safety of the trees it became impossible for the female to establish anoestrus and so they (axiomatically) became permanently receptive.

Once permanently receptive, it is not only unnecessary but also dangerous to come on heat. Any return to oestrus would attract all the males in the area and disrupt the tolerable levels of mating that she hoped to achieve by bonding and being defended by one (or a few) group males. Coming into oestrus would increase male-on-male violence as the established males defended their rights to the permanently receptive females against those males attracted by the signs of heat. Thus, selection has ensured that the overt signs of oestrus in the female hominins have become masked.

A permanently receptive female, faced with the power of the male sexual drive, will benefit from any respite and those females who did not come on heat and advertise their condition, or even their existence if hiding in caves or woods, would have a selective advantage over their less fortunate sisters. For if the permanently receptive female cannot find respite from the male then she would be overwhelmed and quickly become too weak to breed.

Permanent receptivity can be understood as a subdued continuous quasi-oestrus where the female should be prepared to comply with the male demands for coitus at almost any time. This is not a simple increase, compared to animals, in the duration of oestrus; it is the complete loss of the anoestrous period. As by far the larger period of the female animal's life is spent in an anoestrous state, this is a serious loss and it results in the hominin female becoming virtually permanently receptive. Hominins are the only mammalian species known to have abandoned the overt oestrous/anoestrous cycle in this way. This species-defining fact is known but it is virtually ignored in discussions about human sexuality. "Humans, unlike any other species, do not have any obvious external signs to signal oestrous receptivity at ovulation." *New World Encyclopedia – Estrous Activity.*

Thus, the female finds herself in a dangerous and bizarre situation: having lost the ability to defend herself she will endeavour to ameliorate her own position by behaving in ways that reduce her exposure to unsolicited male pursuit. This is difficult to achieve, for once the female is targeted she is desired by another irrespective of her own thoughts, wishes and plans.

The female knows she is going to be forced to mate without the effortless guidance of her oestrous hormones. She has become a victim to the whim and behest of the male and she must marshal her mind and body to deal with this troublingly difficult situation which puts constraints on her mobility and jeopardizes her safety. One of the ways open to her (that has benefits for both females and some males) is to seek the protection of a strong male, either to pair-bond or to be one of a group of females headed by a powerful male (discussed in Chapter 4).

Being permanently receptive means that the female must be available, prepared, and ready to participate in sexual intercourse (as if she was oestrous) whenever a controller (husband, protector or rapist) so chooses. This creates a passive situation in which the hominin female has conceded to a socialization or domestication of her behaviour. An intense pressure selects those females who are compliant and exhibit less or none of the visual, auditory and olfactory stimuli of overt oestrus and this allows them to keep a much lower profile and in so doing improve their fitness for life in a world of motor-control.

Many of the social defences that developed later are dependent on this loss of overt oestrus, but its success creates another problem: from time to time, the female will want (need) to breed and today she must then 'advertise' that fact in other ways, for example by a genuine readiness to lie down (abandon the bipedal position) and/or the removal of clothes. If a permanently receptive female finds herself in a secure society with established rights that give her physical and legal protection then she can 'gift' her sexuality and thereby gain a power that she is able to exploit to her own advantage, and here the arms race between male and female enters its most female-friendly phase (in sharp contrast to those females who are always considered to be the 'property' of the male.)

However, the permanently receptive human female is consciously aware that she still needs to defend her genitalia from unsolicited male intrusions. The animal female does not have this difficult task, so she is not preoccupied with her genitalia throughout her long period of anoestrus, even when she is urinating, or at parturition, it does not become a cause for embarrassment or concealment. Only when the animal is in oestrus does she become more aware of her genitalia but then only so she can position herself to receive the male.

It is impossible to be certain when the hominin females first lost their ability to establish anoestrus, but it is certain that they would have been very vulnerable as soon as they lost the safety of the treetops. We know that chimpanzees exhibit a short overt period of oestrus (the animal norm) and that bonobos have extended their period of oestrus to an untypical length of time [8] but bonobos are not permanently receptive because they can establish anoestrus (avoid mating) at any time by escaping to the treetops. Orangutan females suffer the most aggressive matings of all the primates (Smuts and Smuts, 1993) and can be mated when pregnant, but her solitary behaviour in the forest canopy seems to have ensured that most of her anoestrous period has remained intact. Whereas the terrestrial hominin female out on the savannah has nowhere to escape, so permanent receptivity would have become the norm from the inception of the first acts of coercive motor-control, that is, it probably began in the very early stages of hominin evolution.

The difficulties that result from not being able to defend anoestrus cannot be overstated. Motor-control creates an unresolvable existential predicament for the hominin female that animals do not experience. Animals can readily defend their autonomy up until the moment they die but the permanently receptive hominin female needs to negotiate all her behaviour with others (particularly her sexual behaviour). To do this relatively safely she requires a greater top-down mental capacity to organize some sort of moral code or social contract capable of providing her with assistance and protection. None of which is required in the animal world because so few females are vulnerable to being mated when they are anoestrus.

It has required some remarkable anatomical, physiological, behavioural, social and cultural adaptations (for both male and female) for the terrestrial hominin females to regain a tolerably adequate protection from the male. These adaptations (discussed in the following chapters) can best be understood in the context of an exceptionally intense arms race between female and male hominins in which ultimately, they have both lost autonomy. The vulnerable female (unable to find true safety and establish anoestrus) continuously seeks to improve her defences and the male continuously seeks ways to override those defences.

I suggest that this ploy and counter ploy started at the arboreal/terrestrial transition and continues to the present day. The escalating nature of this arms race creates a heavy maintenance cost for both parties and it also creates an extremely intense pressure point of selection on the female line, for any aspect of anatomy, physiology or behaviour that provides respite for the female is bound to be reflected in the survival rates of her progeny. In spite of the selection of these ongoing defences, none have yet been able to restore to the hominin female the safety of anoestrus, and so it is unlikely that she will, ever again, be able to experience the total safety and protection that nature has afforded to virtually all other female animals.

To be the only species to have abandoned the system of the oestrous/anoestrous cycles that have served all other amphibian, reptilian and mammalian species so well for over 350 million years will take some understanding, for it means that something quite extraordinary has befallen us. It is hard to overestimate the consequences of this change for it is the underlying drive of hominin–human evolution. *Homo sapiens* have emerged out of this upheaval with our present behavioural patterns but we have not yet seriously questioned why our behaviour is so anomalous. It is only by teasing out the reasons for the differences between human and animal sexuality that we can come to a full understanding of the human condition.

This chapter has described some of the forces that could have given rise to the cataclysmic series of events that occurred at the primate–hominin transition which went on to shape hominin evolution; they will be referred to throughout the rest of this book. They are:

1. The use of hands to grasp, hold, motor-control and coerce
2. The female's inability to defend anoestrus
3. Coercive mating
4. Permanent receptivity
5. Loss of overt oestrus
6. Loss of autonomy
7. Heteronomous controls.

I suggest it is this core group of events that has given rise to the anatomical, physiological, behavioural and cultural adaptations that make up human evolution. There is an unresolved dynamic between them that creates a permanent tension that is discharged in *Homo sapiens'* restless energy.

Over time, other major anatomical and physiological changes would have been selected that improved female protection and survival rates and these adaptations are discussed in the next chapter.

Endnotes: Chapter Two

Endnote 1

It is thought that the date of divergence of the hominin line from a close relative of the chimpanzee took place about 4–7 million years ago. However, the important date for the purposes of this paper is the first use of hands to motor-control.

As the control itself leaves no fossil record it is only possible to draw inferences from the changes in anatomy that have appeared that may have resulted from problems associated with motor-control. This date may be the same as the ape/hominin divergence, but it is also possible that this speciation simply created another chimpanzee-like animal and that not until later (perhaps much later, say, 2–3 million years ago) did motor-control take place to the degree required to produce the changes that eventually led to mankind. However, as the first hominins that emerged from the transition showed changes to the anatomy of their lower limbs and feet that made them capable of both arboreal locomotion and bipedal terrestrial locomotion, and the fact that they deviated from the long-established gait of knuckle-walking, would tend to indicate that at this early stage they were already starting to experience problems associated with motor-control.

Whatever the reasons for coming down from the trees (loss of trees, seeking new food sources) and whatever the date (early or late) and whatever the site (Africa, Europe or any other continent), once the protohominins were forced to adopt a mainly terrestrial lifestyle they would have had difficulty defending autonomy. In fluctuating climate conditions there would have been periods of regrowth, where the trees regained an earlier density. However, it would only have taken one period of terrestrial living to develop the fateful practice of the using of hands to motor-control and once learnt by the male he is unlikely to relinquish this dark art, even if the females regained some degree of safety during times of reforestation.

Irrespective of a more accurate chronology of the repeated loss and renewal of forest area it is unlikely that hominin evolution was a steady, gradual process. There would have been periods of expansion and periods of decline linked not only to climatic changes but also to fluctuations in the relative success of the new anatomical and behavioural defences that resulted from the arms race between the hominin male and female.

I suggest it is likely that motor-control was a problem from the start of the ape/hominin divergence and over the next 5 – 6 million years many separate lines developed and faltered until we are now only left with *Homo sapiens*, the one hominin that adapted to the difficulties and violence that followed in the wake of motor-control.

Endnote 2

The term 'intraspecific' needs expansion. Predatory animals hunt and kill members of other species (interspecific) so they become a danger to their prey and this creates, in the prey animal, the specific defence of trying to achieve total avoidance of the predator.

For the victim of intraspecific attacks, the problem is more difficult to deal with because it is not possible to totally avoid members of one's own species since they provide support and protection from other dangers and social needs. But once the hominins learnt to motor-control their own species by holding them, directing their behaviour and coercively mating (and on rare occasions cannibalising them for food), the patterns of a sociable family society were severely disrupted. In evolutionary terms, finding adequate defences to these intraspecific controls have proved to be very difficult, if not impossible.

Endnote 3

Most animals have hooves or pads to protect their feet. Hooves cannot be used for any kind of manipulation, paws are more adaptable because they have claws, and these can be used (with the aid of the mouth and teeth) to catch, manipulate and eat the prey. Animals cannot use their paws to hold and manipulate their conspecifics to any significant extent without inflicting life-threatening damage; whereas, a hominin victim can be motor-controlled by the hands of a conspecific without suffering lasting physical damage.

Monkeys and primates have arms and hands and on occasions can walk upright when their hands are temporarily freed from locomotion and the orangutans in addition have prehensile feet. Monkeys have hands, but they lack strong opposable thumbs and they do not have shoulders, elbows and wrists that are designed for brachiation (an essential feature for motor-control), so if a conspecific is held it is relatively easy for them to break away by forcing the arm joints of the controller into painful positions, thereby overcoming any potential danger.

Primates such as gorillas, orangutans and chimpanzees are capable of brachiation and they do have opposable thumbs, to varied degrees, but they are not as dexterous as the hominins. They can hold and motor-control their conspecifics and they do so on occasion. This is not a regular or expected occurrence (as it is with the hominins) because a potential victim can easily skip away and escape by reaching a position of safety high in the treetops. Nevertheless, the great apes should be thought of as representing an intermediate group that lies between the quadruped mammals (who never motor-control because they do not have hands) and the bipedal hominins (who can motor-control virtually at will). (See 7.)

Endnote 4

By using two hands to motor-control, a significantly different hold (grip) is created to that achieved by using the mouth. The arms allow for movement and for creating distance in the hold, indeed one hand can hold whilst the other adjusts the position of the victim (controlee) who is not necessarily seriously physically hurt. Being held by means of a mouth brings the struggling victim (prey) into extremely close contact with sharp dangerous teeth and it means that the hold cannot be easily adjusted so the victim is usually injured or killed and carried away to be eaten.

In non-adversarial circumstances the mouth can be used as a gentle and delicate instrument. In certain species, for example crocodiles, the mother can use her mouth to enclose her offspring and carry them off to safety in the first few weeks of their life. Other species such as cats and dogs can use their mouth to hold the area of loose skin at the back of the neck, so the offspring can be picked up and carried to safety. While being transported the offspring suspends autonomy (does not resist) and in this way, is not harmed. This type of retrieval does not take place post weaning.

Endnote 5

Endnote 5 covers several interrelated topics, which include 'The safety/danger scales', 'Motivation' and 'Decision-making'. It amounts to an exceedingly brief outline of the underlying operating principles of animal behaviour showing how it has been subject to a major modification in the hominins to deal with the many difficulties presented by motor-control.

The range of movements available to arms, legs, trunk and head are somewhat limited, and the eyes, ears, nose, mouth and tongue are even more restricted. Nevertheless, the brain can organize these small movements into complex patterns which satisfy the essential body needs for food, water, reproduction, avoidance of danger and attraction to safety. The brain not only coordinates and implements these complex behavioural patterns, it does so at the appropriate time, and this is the key to understanding autonomy.

Motivational Systems:

As the first motile life forms evolved, many new responses were created out of the interplay between motility and the selection pressures present in the environment. These pressures, initially acting on the random movements of the organisms, would have quickly selected those organisms that moved towards safety (or away from danger) and this simple orientation is the basis of all behaviour. But motile organisms experience danger from several different sources, for example, lack of nutrients, temperature extremes, predators, and difficulties in reproduction. Evolution selects methods of detecting and responding to each danger, and groups them into what have become distinct motivational systems, see R.C. Bolles (1967), F.M. Toates (1980), D.J. McFarland (1985).

There are a limited number of systems that motivate all mammalian behaviour and they include the following:

1. Energy input
2. Water balance
3. Respiration
4. Thermo regulation
5. Basic body ease (postural shifts, grooming, and waste elimination)
6. Maintenance of balance and coordination
7. Reproduction (mating, birth, maternal care)
8. Social cohesion
9. Regulation of activity
10. Avoidance of external dangers (including maintenance of individual space)

Each motivating system represents a fundamental requirement necessary for the existence of an individual motile life form. These systems differentiated early during motile evolution and they are common to all creatures. Whereas the response programmes associated with the motivating systems, such as those necessary for specific diets, different types of locomotion, varying degrees of territoriality, or methods of reproduction, have arisen because of environmental pressures acting upon the motile forms, so they are widely variable between species.

The behavioural expressions of the motivating systems are easily observed, for example, eating, drinking, balancing, looking, or resting. What is not so obvious is the fact that there must be a decision-making process underlying these activities so that an appropriate response meets a stimulus at the appropriate time.

Decision-making and the safety/danger scales:

Animals are incapable of expressing all elements of their behavioural repertoire simultaneously, yet each motivating system relates to an activity that is vital for the survival of the individual so, in all but the simplest organisms, competition between motivating systems must occur. This has resulted in the evolution of a priority system which ensures that, at any one time, the stimuli representing the greatest threat to the individual will always meet with a suitable response. For example, at the simplest level, a hungry animal will stop eating and flee as the predator approaches. If the animal survives it will recommence eating when the danger has passed and the need for energy inputs, once again, represents the greatest threat to its existence.

For these decisions to be efficiently executed there must be a neural circuit that is able to switch between responses giving priority to the motivating system that defends the individual from the most urgent threat. This neural circuit can be understood as the primary instinct of autonomy, for it is unfailingly able to select a response to the greatest danger and defend autonomy bottom-up by switching from one motivating system to another. However, to do this it must have access to all the information concerning the individual's internal bodily states, motivating drives and external stimuli. As this information has become concentrated by selection pressures into degrees of safety or danger, I have called the form in which it is organized and presented 'the safety/danger scale'.

Broadly, there is a separate safety/danger scale for each motivating system, and each scale has many associated responses which may be triggered by stimuli from the body, the brain, or by external sources, or a combination of all three.

Different inputs are used to set the level or height of each separate safety/danger scale. Scales concerned with the maintenance of body function are based on feedback from internal body organs, such as energy requirements, hydration, and temperature. The reproduction scale is based on the cyclical patterns of hormone levels, and the external danger scale is based on levels of anxiety resulting from the anticipation of physical damage and pain. The safety/danger scales are more complex than a simple linear scale. For example, the energy-input scale is not only concerned with calories, it must also monitor and balance hunger and satiety in relation to several nutrient inputs. Similarly, the female reproduction scale must regulate all parts of the cycle. Nevertheless, the many facets of each scale are integrated and coordinated so that, broadly, a single safety/danger scale can be said to operate for each motivational system.

The way that a safety/danger scale responds to an external stimulus depends upon the height of the scale when the input is received. This determines the relevancy, or 'weight', of the information. For example, a hungry tiger will stalk a deer that a sated tiger would ignore; or a female bird will respond to a male in the breeding season yet ignore him the rest of the year.

Evolution ensures that the calibration of the safety/danger scales interrelate, so that the level or height of each scale can be continuously monitored and used to determine which response gains the priority to become the next unit of motor activity. The sated tiger, above, may not be moved to catch the deer, but in the midday sun, it would be moved to find shade. Thus, the highest scale (not necessarily at its highest point) has priority because it represents the most threatening danger at that time.

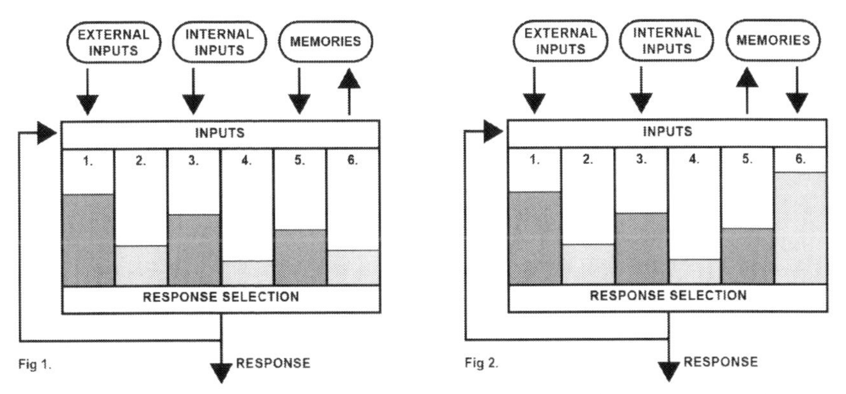

In fig 1. the safety/danger scale 1 is set higher than the other scales and so feeding is given priority. In fig 2. external danger scale 6 is the highest scale because (say) a predator has appeared and so now it is granted priority.

The principle of giving priority to the highest scale not only defends autonomy from all external threats, it also attends to all internal needs, and so it achieves homoeostasis by maintaining the scales at, or nearest to, their optimum (safe) levels.

Direct competition between motivating systems has been refined during evolution, for example, the individual does not become bogged down by making small changes that would result from the rapid back and forth switching from one system to another, say, a sip of water for every morsel of food. This inefficiency threatens survival, so it would have been selected against by building time constraints into these essential, but non-urgent, responses.

Also, the context in which a stimulus is received can weight that stimulus, for example, the last water hole for miles will attract an animal that is not exceptionally thirsty. The sight of the water hole triggers an anticipatory threat of serious thirst later in the day, and this combination of stimulus plus context allows the water balance motivational system to gain priority.

"In demanding environments evolution is bound to select improvements that take account of incipient activity", said McFarland D.J. (1985), but the fact that the competition is 'refined', rather than 'simple', does not invalidate the principle that competition between motivating systems is the basis of animal decision-making. This bottom-up decision-making system is the primary decision-making system of all animals and it is equipped, from birth or soon after, with adequate responses to virtually all eventualities an individual of that species is likely to meet, which are then updated by experience. This gives the individual lifelong bottom-up protection from all the established dangers in its environmental niche without the individual animal being responsible for creating, choosing or updating those responses.

With the advent of motor-control, a new danger emerged for the hominin: conspecifics posed a serious threat to autonomy and the external safety/danger scale developed to respond to this new danger. In so doing it has undergone modification and enlargement (encephalization) to create a top-down secondary decision-making system that is able to comply with the demands and vicissitudes of heteronomy.

The hominin external danger scale (6 above) has been expanded to deal with the new and ever-present threat to autonomy that comes from motor-control and heteronomy. This threat is so serious that it readily gains a top-down priority over virtually all the responses of the other safety/danger scales, thus a high percentage

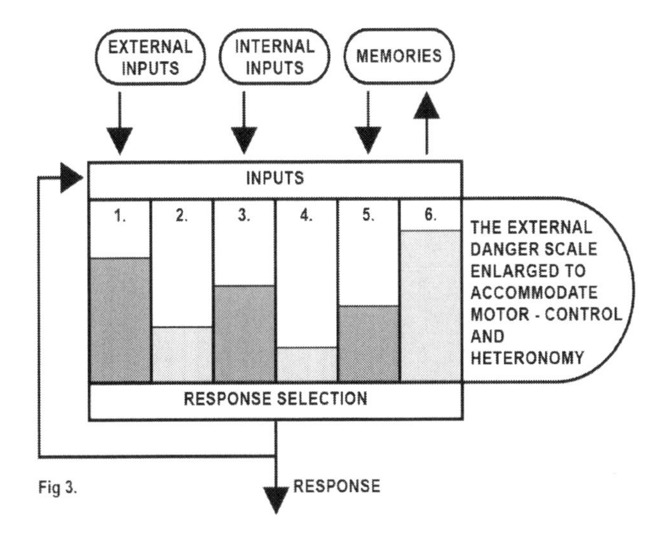

Fig. 3

of human activity is concerned with defences to, and compliance with, the everyday heteronomous control detailed throughout this book.

Thus, before any hominin response is granted priority (even those that would have smoothly and autonomically gained priority in the animal, such as urinating and defecating), it must be carefully monitored by the hominin's external danger (motor-control) scale for any element that has the potential to disturb the uneasy balance of heteronomous governance. Thus, the motor-control element of the safety/danger scale gained a top-down power of veto and an ability to modify the responses of the other scales which has resulted in hominin behaviour becoming considerably more flexible and adaptable than that of other animals. All of which requires a much-increased cerebral capacity.

Endnote 6

The time when an individual establishes full autonomy varies with different species. Some, such as bison and deer, have precocial offspring that are fully autonomous from birth, others such as hedgehogs, cats and dogs have altricial offspring that have a short period of dependency before weaning, during which their autonomy can be suspended while they are carried to safety (retrieval) by a parent. In this situation the pup or kitten will suspend autonomy (implicitly trusting their dam) and become motionless, thereby avoiding physical damage and any conflict of governance that would otherwise ensue from being moved around (motor-controlled) by others.

Nevertheless, individuals of all species, except *Homo sapiens*, are independent by the time they are weaned, or soon after, and they never lose autonomy except at death or in their contacts with mankind. This includes all other primates who are meticulous in their distinction between allowing their young to 'hitch a ride' (preserving autonomy) and 'carrying them off' (overriding autonomy). This meticulous preservation of autonomy also applies to feeding; before they are weaned the young of some species, such as birds, will solicit food and they will be fed by their parents but they are not fed against their will (unlike some human children who, as a punishment, have been made to eat their own vomit or drink a pint of piss a day, see *The History of Childhood*, edited by Lloyd de Mause.) Post-weaning animals never feed (that is put food or a teat in the mouth of) their own, or another's offspring, and this ensures that the offspring, must always feed itself and this rule applies even in sickness. When ill, you do not lose autonomy, you simply are less able to protect it, but you go on protecting your own autonomy (with minimal assistance) as best you can, until you die. These seemingly harsh rules reflect how important the principle of the preservation of autonomy is within the animal world.

Endnote 7

Coercive mating in non-primate animals:

Coercive mating has been reported in lower invertebrates, insects, fish, amphibians, birds, primates and other mammals (see Shields & Shields, 1983). The incidence is rare and many (possibly most) of these cases are disputed as not being examples of truly coercive mating ('true rape') (Estep & Bruce 1981, Gowaty 1982, Kitcher 1985, Smuts and Smuts 1993, Clutton-Brock and Parker 1995). Each illustrates different facets of the interplay between female choice, motor-control, and the oestrous cycle in situations where the anatomical and environmental constraints on the male are absent or reduced.

1. The primates that are known to suffer coercive mating are the orangutan and the common chimpanzee and this is discussed in Endnote 8.
2. The insects are *Drosophila* (Manning 1967), and scorpion flies (*Panorpa*)

(Thornhill 1980b). Male scorpion flies have muscular genital clasps and a clamp-like notal organ on the abdomen with which they can catch and hold the female. The males have three mating techniques, one involving force, in which the male grabs and mates a passing female without the usual nuptial offerings, or pheromone attractant. It is having a clasp and a clamp (like two hands) that produces the unusual dexterity, allowing the male to hold the female while he positions himself for mating.

3. The birds are waterfowl, the lesser snow goose, and the mallard, shoveler and pintail ducks (Barash 1977; McKinney, Derrickson & Mineau, 1983). With access to land, water or air, the female waterfowl could, with a little more caution, easily defend herself, especially the duck, as she appears to run the risk of being drowned when rape is attempted by several drakes (Huntingford & Turner 1987). However, it should be noted that, as soon as incubation begins, geese or ducks are not molested by the male, even though they are physically vulnerable; presumably this is because incubation signals the end of the mating period.
4. The amphibians are frogs (*Rana*) and toads (*Bufo*) (Wells 1977, Howard 1978). When mating, the male rides on the back of the female fertilizing her eggs as she spawns, his grip being facilitated by forelimbs that have nuptial pads. The element of choice the female can exercise in this situation is difficult to establish, however, she is in oestrus in the sense that she sheds ovum to be fertilized by the male, and if she is coercively mated (clasped by a male not of her choice) then it is because the male is able to grip her with his specially adapted forelimbs.
5. The fish 'rape' is not forced copulation but a violation that may be said to occur when the female deposits her eggs and they are fertilized by an intruder rather than the chosen male (Barlow 1967). Although the female's choice of male has been overridden here, she has not really lost autonomy.
6. The non-primate mammals that may experience coercive mating are elephant seal (*Mirounga*) cows who vocalize loudly when they mate. They vocalize (protest?) more loudly when mounted by a less dominant bull, which alerts the dominant male who then drives off the intruder (Cox & LeBoeuf 1977). If this is rape, then it occurs because the female is trapped (out of water) on the beach breeding grounds. Also, a lioness may suffer a manipulation of her oestrous period; this is not coercive mating, but it is an example of the male sexual drive having power over the female. When

a new alpha male lion takes over the pride he will kill all the young lions that are still suckling. This brings the lioness into oestrus which allows him to mate and produce his own progeny (an outcome, presumably, selected by evolution because it facilitates social stability). The breeding females (unlike the dead cubs) do not suffer any loss of autonomy, they are not held or coerced into anything, they are simply unable (or unwilling) to defend their cubs and subsequently react to events by coming on heat.

In none of these examples of coercive mating is the female as defenceless as the hominin out on the open savannah, because the animal female can establish some sort of anoestrus when she is released. However, they do illustrate how vulnerable the oestrous female can become if the males can grasp and the female is unable to fully exploit the environmental constraints that would otherwise restrain the male.

Endnote 8

Coercive mating in Primates:

The current scientific views on sexual coercion in primates are to be found in:

Male Aggression and Sexual Coercion of Females in Nonhuman Primates and other Mammals: Evidence and Theoretical Implications, by Barbara Smuts and Robert W Smuts in, *Advances in the Study of Behaviour*, vol. 22, 1993. And, *Sexual Coercion in Primates and Humans: An Evolutionary Perspective on Male Aggression against Females,* edited by Martin N. Muller and Richard W. Wrangham.

The great apes can be thought of as an intermediate group between the quadrupeds and the hominins because they have hands that, to varying degrees, can be used to hold others, so potentially they do face difficulties that are not experienced by the quadrupeds, as can be seen by the fact that they do rely on the treetops to help establish safety and maintain the anoestrous phase of their cycle. In keeping with an 'intermediate species', the female *common* chimpanzee and the orang-utan do on occasions experience what are considered coercive matings. But, unlike the hominins, all have retained their oestrous/anoestrous cycles in some form or another, either by extending the oestrous phase or by tolerating some mating in the anoestrous phase.

Chimpanzee (*Pan troglodytes*): Forced copulation rarely occurs in chimpanzees, but the females can be subjected to serious *physical* male aggression. Lone, low-status females on occasions may be attacked, battered or even killed if caught by a group of marauding males. On rare occasions old, low-status males may also be attacked and killed. Also, a group of chimpanzees will hunt, catch, hold, kill and eat an individual from another group; this cannibalism is dependent upon the use of hands to immobilize the victim. It does not appear that these episodes of holding and killing have mutated into coercive mating. Chimpanzees occasionally hunt monkeys and they do so in groups. The victim is chased, caught, held, and with canine teeth, killed. This is standard omnivore predation except here the hands are used to catch and hold (rather than the mouth and claws).

Orangutan (*Pongo pygmaeus*) females experience the most forceful (aggressive, coercive) matings of any non-hominin animal. They are vulnerable to being caught and coercively mated by solitary males who, with their grasping feet, have even more prehensile appendages than hominins (MacKinnon 1974; Galdikas 1975, Muller and Wrangham 2009, Smuts and Smuts 1993). This also includes being mated when they are pregnant and so their period of anoestrus is not guaranteed but, unlike the hominins, the orang-utan female can find some safety in the treetops. Even so, her vulnerability, and her month-long oestrous receptivity clearly shows that orangutans also fall into the category of an intermediary species between the hominins and the quadrupedal mammals. Orangutans do not kill members of their own species. They have the longest childhood dependency of all the apes, nursing their young for six years, and the females give birth about every eight years. The serious loss of habitat, due to forest clearance that orangutans are presently suffering, is likely to make this female vulnerability much worse.

For Gorilla (*Gorilla gorilla*) females, coercive mating appears to be an extremely rare event presumably because the males are much heavier and so the female can always gain safety in the treetops. (Other possible reasons for this degree of female safety are the shorter length of the opposable thumb and/or the dominance of the silverback male who regulates most of the mating opportunities or a general lack of intraspecific aggression.)

Bonobo chimps *(Pan paniscus)* can be seen to deviate from the basic sexual norms of most animals, since their oestrous period has been extended to include the period of pregnancy and suckling in which they tolerate mating (that is, they do not have a firmly established period of anoestrus). They are promiscuously sexually active, but they are not classed as being permanently receptive because their semi-arboreal lifestyle means that if the female is seriously distressed she can always retreat to the trees for safety.

However, tension in the bonobo is usually resolved by offering more sexual contact rather than withholding it. Despite the overtones of pleasure, bonobo male–female relationships seem to be more agitated/troubled/disturbed, compared to, say, those of the gorilla (or quadrupedal animals) and their behaviour can be understood as falling into an intermediate position between humans, who cannot establish anoestrus, and all other animals that can easily establish anoestrus. So, it appears from their behaviour that bonobo females do have an underlying difficulty establishing anoestrus which has resulted in them extending their oestrous period and increasing their willingness take part in sexual activity to placate the male sexual drive. (Note this has moved into areas beyond reproduction to include adult and young male–male and female–female sexual contact, which implies that the background tension of intraspecific motor-control drives both sexes to (it has been said) 'make love not war'. This appears to be a successful tactic because there are no reports (at present) of possible incidents of bonobo coercive matings.

Female bonobos spend more time together than do most female chimps and they have a greater influence on what their group does. This female solidarity seems to temper the worst excesses of the underlying violence that persist in the common chimpanzee male-dominated hierarchies.

The degree of sexual activity undertaken by the bonobo must have considerable cost in terms of time, disease risk, and inattention to danger. Given that, for good reasons, so many other animal species have reduced the amount of time spent having sexual contact to a minimum, one must look for the reasons for the bonobo's excessive sexual activity. Seeing their sexual behaviour as a way of dealing with problems like, but not the same as, those faced by humans, illustrates the difficulties both species have living an autonomous life in the presence of hands that can grasp, hold and control.

Baboons (*Papio hamadryas*). Of all the monkeys, baboons are genetically the most closely related to humans. Baboons are vexatious and at times will use their hands to steal the recently born young of other baboons. In some troupes this can become so disruptive that the social cohesion of the group completely breaks down. Baboons have hands that facilitate mischief, but they do not have brachiating shoulders or opposable thumbs, so they are unable to grasp, hold and manipulate larger resisting individuals to the extent of, say, the chimpanzee or hominin. In fact, mating for the baboon female is a gentle process with the male grooming the female and bringing her gifts of food.

CHAPTER THREE

DEFENCES TO MOTOR-CONTROL, ANATOMICAL AND PHYSIOLOGICAL ADAPTATATIONS

We can now start to view hominin evolution from a new perspective, that of seeing many of the specifically hominin adaptations that have arisen as resulting from the pressures of motor-control and heteronomy. In this chapter I will consider the anatomical and physiological adaptations that are most likely to have been influenced by these new pressures.

Some of the anatomical adaptations that became important in the overall story of human evolution were in place before the protohominins began to evolve. The great apes had specialized in arboreal locomotion and this ability was dependant on the grip of the hand with its opposable thumb, in combination with the brachiating multi-joint flexibility of the shoulder, elbow and wrist, which enabled the apes to swing with ease through the forest canopy. At the arboreal/terrestrial transition these features became redundant for arboreal locomotion but were nevertheless inherited by the terrestrial protohominins, who found that they were ideally suited for grasping, holding and coercing their fellow protohominins.

Following the protohominins' descent from the trees several new, specifically hominin, adaptations were selected such as bipedalism, loss of hair cover and encephalization. The current evolutionary explanations for these changes are mainly variations of a theme based on the difficulties of finding food and avoiding predators out on the open savannah, (dangers, it should be noted, that are common to most mammals). These selection pressures are discussed at length in text books of human evolution but, I suggest, they are not a sufficient or satisfactory explanation, primarily because the hominins have evolved in ways that are quite different to all other animals (hence humans do not like to be classed as an animal), and so we need to explain why the hominins have taken this unusual route and why these differences have evolved.

These specifically hominin features (anatomical, physiological and behavioural) are extant so they must have satisfied the basic evolutionary criteria of improving fitness. The question is, what fitness do they improve upon? We need to look at them in the light of the difficulties that motor-control and sexual coercion created for our ancestors.

Being caught, held, heteronomously controlled and coercively mated are all traumatic, dangerous events that would have exacted a heavy toll on the early hominins. Humans have become inured to this gross trespass, so we are insensitive to the depth of the trauma and the damage that the loss of autonomy originally inflicted upon our animal nature. I suggest that motor-control and coercive mating was initially so debilitating that it would have severely reduced the female's ability to breed and nurture her offspring successfully.

Set against this harsh background any behaviour or anatomical change that gave rise to a respite in the experience of these controls would have been selected. Anything that reduced the worst excesses of the control and coercion would have improved the fitness of the controlee and increased the female's ability to breed successfully.

The damage done to the female in coercive mating is not just the physical act of sexual penetration, it is also the continuous pursuit, hounding and manhandling, from which she has no respite. This harassment results in a prolonged loss of autonomy and gives rise to the perpetual fear of its reoccurrence, and it exacts a heavy toll on the victim in addition to any physical injuries and the restriction of feeding opportunities that she suffers. The hominin female must have also lost the 'mental equanimity' that she originally shared with the animals. This equanimity stems from the smooth functioning of the bottom-up programmes (that is lost with coercion) and it provides the contented base that is so conducive to the ability to breed successfully (see Chapter 13).

The nub of the argument here is that although coercive mating does occur in some other species it has not reached the unrelenting levels that occurred in the hominins when the female lost her ability to establish anoestrus and it is this persistence that necessitated the selection of new behaviour, physiology and anatomy that facilitated new defences and the complex ability to comply with the heteronomous instructions of the controller.

This new hypothesis allows us to examine the specific hominin adaptations in the light of the idea that defence to motor-control could have provided the full, or partial, driving force for their selection.

3:1 Bipedalism:

Bipedalism, walking upright on two legs, is a method of mammalian locomotion unique to the hominins and is an early hominin adaptation certainly 3–5 million (possibly 5–6) million years old. However, there is a high cost to bipedalism, so a powerful pressure must be operating for this adaptation to have been selected. Walking upright is structurally weaker than the knuckle-walking of the great apes, tripping, falling and lameness is more of a handicap to a biped than it is to a quadruped and the upright posture exposes the male genitalia to injury (whereas the genitalia of a male quadruped are significantly better protected). In addition, it is accepted that bipedalism is twice as energetically expensive as quadrupedalism. Although it is claimed that its efficiency lies in the ability to run down prey over long distances and/or to be able to look over tall grasses to see a prey or predator, it is hard to believe these had sufficient benefit for an efficient omnivore to adopt such an unusual and costly posture. Clearly, a change in the method of locomotion was required as the protohominins came down from the trees but their existing knuckle-walking could have become more efficient or they could have adopted a quadrupedal gait; after all, this is the standard design for savannah-dwelling mammals be they herbivore, carnivore or omnivore.

Now consider the advantages of bipedalism to the protohominin female who is unable to establish anoestrus.

3:2 Bipedalism and mating, the female perspective:

At the beginning of the protohominin's terrestrial existence there were very few defences immediately available to the knuckle-walking female who, trapped on the open savannah, was chased, caught, held and coercively mated.

Rearing is an obvious first response. The female rears up, dislodges the male from her back and in so doing displays greater height and this, together with the effect of novelty, frightens off the male or at least distracts his sexual intent. In an upright position, the female turning circle is reduced, she can face her assailant more quickly and her hands are free to scratch and push away. Initially, both male and female would be unsteady in the standing position and this would be to the female's advantage (even when the male is bipedally competent he continues to be disadvantaged compared to the knuckle-walking male holding the female in a rear-mounting position).

When the female is upright her genitals are far less accessible which means that the male is unable to mate without some cooperative movements from the female, or unless he resorts to greater violence and positions her against her will. Thus, rearing would be a very effective defence for the female and any return to knuckle-walking would benefit the male, so the fully bipedal mode of locomotion would tend to be selected because of the increased protection it affords the female.

When the bipedal hominin female faces her attacker, she can always bend her lower body away from the male and generally prevent or dislodge any penile intromission. Thus, if she can maintain a face-to-face bipedal position, some cooperative movements are still required even if she is held by the male. However, cooperation is not required if she is backed up against a tree or rock face that restricts her flexion, nor when the reluctant female is forced to the ground and mated in the supine or the ventro-dorsal ('doggy') position.

So far, we have considered attacks from a single hominin male but if several males cooperate with each other the female's situation becomes even more desperate. A terrestrial hominin female only has to be placed in a headlock by one male for her to become totally vulnerable to the other males. Thus, she will try to hold her head at its maximum height so that the risk is reduced. In contrast, animal females can always establish anoestrus even when in the presence of many males; for even if the males cooperate, they can only corral the female, they cannot hold her in a position that sufficiently restricts her movement to allow penetration. Only when the female animal restricts her own movement and 'stands on heat' can mating take place.

Thus, in the hominin female's new need for protection

from male advances, bipedalism does improve her defences and it also carries with it the visual statement that reads, "I am less 'available' than my knuckle-walking sisters or cousins." The number of movements required of the bipedal female to adjust her body posture for mating to take place is far greater than those undertaken by the quadruped female, who simply needs to stand still. So, if the female can adopt and maintain the bipedal position the level of mating difficulty for the male is certainly increased.

Bipedalism establishes a situation where the females are now required to exhibit a minor degree of cooperation for mating to take place. The hominin female, having lost her ability to establish anoestrus, is left with this smallest degree of control – set against a background of overwhelming violence, she is now able to position herself to assist, or thwart, the male desire. This is a very small gain but at this early stage of the 'arms race' between male and female, it is all that is left of her autonomous control after her own oestrous behaviour had been suppressed. Being vulnerable to physical control, injury or punishment means that whenever the female does enter an area of close contact with the male, she will always experience, to a greater or lesser extent, an unresolved approach/avoidance conflict.

Eventually, the males would develop ways to overwhelm the bipedal female and (like all other mammals in seriously threatening circumstances) she would have to submit. In submitting, the hominin female falls to the ground and lies with her vulnerable belly exposed. Initially this would have unsettled and disadvantaged the male for it is extremely unusual for an intraspecific attack by an animal to continue past the point of submission, especially when the victim is a prized breeding female, but attacking past the point of submission has become a defining feature of human nature. Also, mating on the ground is a new experience, requiring a new body posture, and this too would initially deter the males. Clearly, these early disadvantages for the male were also eventually overcome for now the female is often flung or wrestled to the ground (or she falls to the ground knowing that she is overwhelmed) and is then readily mated in the supine ('missionary') position, bizarrely taking the weight of the male on her vulnerable abdomen rather than on her strong backbone (as is the case with ventro-dorsal matings of most other mammals).

Bipedalism leads to an increased use of the sitting posture. Sitting is a more upright (alert) position than lying down and when threatened it allows the individual to rise more quickly, facilitating a very quick transfer to bipedalism. Sitting is a convenient mode of rest for a tailless biped and fleshy buttocks make sitting a more comfortable experience. Sitting conceals the female's genitals and makes them less accessible and this can be exploited by the females in a safe group environment (see Chapter 4) for a determination to sit cross-legged (an unusual or impossible pose for other species) may be the nearest a permanently receptive hominin female gets to being able to establish a short period of quasi-anoestrus.

In addition to sitting, fleshy buttocks also benefit the bipedal female in an upright posture for their bulk adds to the concealment of the genitals, making them less accessible to the male, which in turn has probably lead to an 'arms race' increase in the length of the penis. Also, the hominin penis has lost the highly sensitive penile spines, common to many other apes, which cause them to ejaculate very quickly. With longer periods of intercourse guaranteed to the hominin male by control and pair-bonding, this sensitivity became redundant.

The genitalia of the quadruped female animal are usually protected by the flexible tail, but they are nevertheless accessible to the male who has a daily need to monitor the female's hormonal condition. In contrast, bipedalism places the tailless hominin female's genitalia in the most inaccessible area of her body making 'inspection' by the male much more difficult. This inaccessibility confirms the situation that, with permanent receptivity and loss of oestrus, there are fewer significant female sexual changes for the male to monitor. In addition to all of this, the wearing of clothes (see Chapter 4) has added extra layers of inaccessibility.

Having lost the safety of being able to establish anoestrus, the hominin female always faces a tension between finding defences to the unrestrained male on the one hand, and her 'need' to breed on the other, which is conveyed to her by means of a silent oestrus. A range of outcomes for the female has persisted throughout history, and these run from having to suffer unrelenting coercive mating, to the current situation of living in a modern Western society where female safety is much improved by an enforceable system of legal and social justice. Individual females need to adapt their sexual strategies to the system that is functioning around them, so it is important to understand the pressures from within her society from which the female must defend herself, for her boldness, or timidity, is dependent upon the support and protection that she can expect to receive from others (see cooperative defences, Chapter 6).

If the bipedal hominin female was lucky enough to find herself in a safe environment she would (when fit and ready) need to signify her willingness to mate and interestingly (given the history) one of the ways she can now do this is by showing a readiness to come down from her bipedal heights, take off her clothes and recline on the floor (or bed).

It is likely that female bipedal rearing came into existence as an act of desperation which went on to provide longer-term benefits. Rearing developed into bipedalism because it afforded the female some respite from the male, compared to her original knuckle-walking posture.

3:3 Bipedalism – the male perspective, mating and fighting:

Once bipedalism proved to be advantageous to the female, it would have been selected and inherited by both male and female alike even though there are some disadvantages for the male. The genitals of the bipedal male are now bizarrely more exposed than those of a quadrupedal or knuckle-walking male, and so they are vulnerable to injury especially when travelling in areas of dense vegetation. In addition, bipedal sexual advances expose the male genitals to a greater risk of intentional, or unintentional, damage by the female than occurs with knuckle-walking, and bipedalism also exposes the genitals to serious damage in male-on-male fights (hence the prohibition in contact sports of hitting below the belt).

There is an additional development that comes with bipedalism (that can disadvantage individual males and females), for once both hands are freed from the locomotor constraints of knuckle-walking they can be used for ever more dexterous techniques of coercion that eventually lead to the use of weapons, ropes, tools, control of fire and to sophisticated methods of restraint, and so hominin coercion (now male-on-male as well as male-on-female) gains further impetuous, with the spoils going to the controller who is not only the strongest but who also has the least empathy.

Male-on-male competition and aggression must also have been affected by the ability to motor-control for if the female could be held past the point of submission so could the male. For primates living an arboreal, or semi-arboreal, lifestyle any injury to their hands or arms would severely restrict, if not end, their life in the trees. So, when fights break out both parties are aware of this vulnerability. However, for the terrestrial, bipedal hominin male, the hands and arms are now less essential, which means greater risks can be taken and more fights chanced. Just as the female is vulnerable to being held and placed in a headlock, so too is the male in intra-male fights.

Bipedalism gets the head up to its highest position away from the fray, so it can also be seen to be part of any defence of the head and throat. Starting a fight in an upright position means that many preliminary skirmishes can be worked through and resolved before the serious holds or injuries occur. The worst outcome is where the victim is wrestled to the ground, held and attacked past the point of submission with the throat becoming particularly vulnerable.

Intraspecific fights between quadrupeds have a standard pattern of circling and biting, or gouging with horns, but the loser is never held nor taken past the point of submission because animals have an inbuilt cut-off point that allows the victim to run away.

Bipedal fights have a quite different pattern, arms flail and both parties stand toe to toe trying to deny a hold that will throw them to the ground. Hominins have probably always fought past the point of submission, humans certainly do, and submission now seems to have lost its (animal) power to end a contest. Indeed, submission may provoke an increase in the ferocity of the attack probably because the victim, if released, can easily be caught and attacked again. This is more like the way a predator 'plays' ('cat and mouse') with its prey but it is not a behaviour pattern that animals have adopted in intraspecific disputes.

3:4 Bipedalism – the child's perspective:

Bipedalism has created several problems for childrearing. The child must be carried before it learns to walk bipedally and this involves a combination of coercive motor-control and solicited motor-control. Also, bipedalism, together with hairlessness, gives rise to a serious added complication (highlighting again that there must have been a major advantage in the adoption of bipedalism for it to be selected) where the offspring must solicit motor-control to be picked up to participate in many vital aspects of behaviour such as being comforted or held to the breast for feeding (see Chapter 8), or being carried when travelling, or to escape from danger. The mother now has an increased responsibility for the child, and so she needs to be sensitive to its solicitations, yet she also must be mindful that carrying the baby is also a control over the baby's movements that has the potential to create a loss of autonomy.

Compared to other mammals, this greatly increases the complexity of the mother–offspring relationship; today a human child may solicit being carried to quite a late age (five or six years old). In contrast, a quadruped's offspring is never carried, and suckling enhances its autonomy for it must find its way to the teat unaided. Primates will carry their young but it is the young that autonomously hold onto the mother, gripping her hair, rather than the mother holding the offspring.

I suggest that bipedalism developed as part of the changes that arose out of the new sexual pressures on the hominin female. For millions of years the male animal's pattern of sexual activity was determined by the female through the activity of her hormones, which produced the changes that brought her into oestrus and prepared her (and the male) for mating. This universal animal pattern was lost when the hominin male became able to grip and hold and bipedalism can be understood as a partial response to this coercion, allowing the female to restore some lost autonomy. The male reacts to this female defence using his hands and arms with even greater dexterity, hence the ramifications of bipedalism on behaviour are profound and ongoing.

Bipedalism is not simply another form of locomotion; it is primarily a vital defence for the female in her need to protect herself from the male. However, it is only a partial defence; it is always vulnerable to the more aggressive males and so additional adaptations emerged over time that further assisted female survival.

3:5 Reduction of hair cover:

> "Another most conspicuous difference between man and the lower animals is the nakedness of his skin."
>
> Darwin, *The Descent of Man*, p.85.

> "Loss of body hair is an extraordinary deviation from the animal norm. Hairlessness is so radical that it is hard to explain. On the African plains hair cover insulates the animal body from solar radiation, but this is beneficial only so long as a satisfactory heat balance is maintained by remaining in the shade, and inactive, during the high, midday, savannah temperatures."
>
> (McNab 1974).

Although hairlessness is a radical and costly solution, leading to increased sweating and water dependence, it has been suggested that in areas of low resources, or where competition with other species is intense, the hominids may have been forced to hunt or scavenge in the midday sun and hair loss reduced the tendency to overheat (Dunbar 1977). Here again it is suggested that it is food shortage that has given rise to major adaptation not required by any other species. However, a much more direct pressure results from motor-control and continuous receptivity, which creates restlessness within the social group, requiring the females to remain alert and responsive to male pursuit, at any time (even in the midday sun). In this situation thermal stress would tend to increase female vulnerability and lead to the selection of hair reduction.

A coincident pressure arises from the ease by which a hairy body could be gripped and held using a tuft growing from any part of the anatomy, whereas a body covered in fine hairs can only be gripped easily at the wrist and ankle. Scalp hair was retained for its benefit as solar protection.

Dr David Read (2011), a parasitologist at the Florida Museum of Natural History, has compared the genetic diversity of head lice and crab lice and has set the date of hairlessness at 3.3 million years ago. This may also date the use of animal skins for clothes, but it is possible that clothes were a later invention.

The retention (and increase in length and thickness) of pubic hair could also have been selected as an early form of concealment of the female genitalia, a need which has been spectacularly superseded by the wearing of clothes.

The combination of hair reduction and bipedalism gives rise to a compounding problem that has required further deviation from the animal norm. The hominin new born young are unable to cling (like other primates) to their mother's hair and move around her body unaided and so the young must be gathered up, and held, before suckling or travelling. This produces a serious loss of independence for the child, who now must solicit motor-control to 'hitch a ride', or to be lifted to find the breast. The necessity for the offspring to solicit motor-control before it can feed is a radical departure from animal behaviour patterns, and will have required modifications to the neural circuits of the child, and the mother, to regulate this dependence. Sensitivity to solicited motor-control places a far greater responsibility on the mother (and others), who now not only have to respond to the solicitation, but also must guard against overriding the child's autonomy by indulging in unsolicited motor-control. A perfect balance is difficult to sustain and so ambivalence is introduced into what should be a close relationship of autonomous individuals. (See also agonistic buffering, Chapter 4.)

3:6 Reduction in the size of canine teeth:

Darwin suggested that the large canines of the male ape are used for fighting and bluff, and that they became redundant when mankind developed weapons. However, it is unlikely that this redundancy is the only factor involved.

Selection against large canine teeth in the hominins could also have arisen because intraspecific motor-control produces close contact struggles in which large canines would inflict serious injuries on the smaller female as she tried to fight off the male. It is one thing to kill an attacker or a rival male, it is quite another for a receptive female, of one's own species, to receive life-threatening wounds while mating. Any situation in which the weaker controlee can be held while the stronger controller inflicts physical injuries would quickly be selected against in the evolution of intraspecific behaviour. In any dispute involving fights with canine teeth, the parties spar, jostle, move back and forth, and either party can, at any time, back off or escape by running away. An intraspecific fight where one participant is held, and the other uses his teeth is too dangerous to be sustainable.

It is not just the teeth; the human jaw has also been refined so that the mouth has lost its fighting function and has increased its range of expressions. These changes can be understood as part of a long line of developments that have 'domesticated' (see Chapters 4 and 6) the hominin, enabling male and female to coexist in a world in which motor-control is an ever-present threat. Reduction of canine teeth could also have been affected by this pressure, for both males and females may now need to smile (Van Hooff 1972) to be able to project a friendly appearance towards those with whom they wish to cooperate.

This selection against the fighting canines has left the alpha males without their natural easy ability to threaten and bluff, to either maintain a stable social order, or to defend the group from external threats. This, together with the vulnerability caused by the threat of motor-control, and the ready ability of the hands to grasp, makes it virtually inevitable that weapon use (see Chapter 4) would be developed for both attack and defence. Hence, the underlying conflict between autonomy and control experiences a further escalation.

3:7 Breasts on non-lactating females:

The mammary glands of all mammals (apart from humans and the high-yielding domestic milking cow) engorge only while the offspring are being suckled. Post weaning, the glands reduce significantly in size, virtually disappearing in many species.

Having permanently enlarged breasts on a non-lactating female is a feature that seems to register as a desirable attribute in the human mind. Yet, having this costly specialist tissue present in a non-functioning state, from puberty, throughout pre-pregnancy, pre-parturition, post weaning and post menopauses is an odd situation that requires an evolutionary explanation.

"It is the premature development of breast tissue before it is needed for lactation that particularly distinguishes humans from other species, although because of this the human breasts are permanently enlarged" (Caro 1987).

There are three hypotheses that attempt to account for full breasts on non-lactating females:

1. Breasts mimic buttocks and are releasers of male sexual behaviour.
2. Breasts hide the female's reproductive condition allowing them to fool their mates and copulate elsewhere.
3. Breasts allow infants to nurse from the hip.

All are inadequate. Breast tissue leaves no fossil record so the dating of the onset of permanent human female breasts is uncertain but clearly, they mimic lactation and act as a powerful visual signal. Lactation is an anovulatory state in which animal females are treated with 'respect' by their conspecifics, they are left alone to tend and rear their young. I suggest the period in which the hominin female gains this advantage of 'care and respect' has been considerably extended simply by maintaining permanent breasts on non-lactating females.

This 'maternalization' of all women (compared to animal females) is part of the general domestication of the hominins. A female with permanently enlarged breasts is less fit to meet the rigours of everyday life; nevertheless, there is a benefit in terms of protection from motor-control because she is initially seen to be lactating and therefore anoestrous and this would make them advantageous in evolutionary terms. At a later stage, well-endowed females may then become prized by the (by now) pair-bonding males, for the growth of large breasts is also a sign of health and nutritional abundance.

3:8 Menstruation and the engorgement of the womb:

Female animals do not menstruate, most of their adult life they are anoestrous, either pregnant or suckling (that is, not in an oestrous cycle). When they do cycle and fail to become pregnant then the lining of their womb is reabsorbed internally exposing a new lining for implantation of the fertilized ovum at the next mating.

The human female's womb at implantation is engorged with nutritive fluids to the degree of other mammals that are two to four weeks into their pregnancy; if conception fails then this lining is of too great a volume to be reabsorbed in time for the next cycle, so it is discharged as the menses. These extra nourishing conditions necessary for implantation in the human female may indicate a difficulty to conceive which is understandable when the female manages the complexities of permanent receptivity against the background of an ever-present threat of motor-control.

The lack of body control (incontinence) of menstruation reduces fitness, restricts mobility, and increases the need for grooming. However, it is possible that this mimicking of a wound also has a selective advantage, since it may protect the female from motor-control in the following ways: male aversion, a temporary lower sexual status for the female, creating a bond and a degree of sympathy between females, and a disruption of the onerous pattern of continuous receptivity with changes in the mood of the female. Remembering that the hominin female is unable to establish anoestrus, menstruating for nearly one fifth of the year (if there is no conception) can significantly improve her defences to male sexual pressure. The great apes menstruate only minimally, no wild animal could survive the selection pressure from predators incurred by the incontinence of menstruation, so it is likely that the heaviness of the menses only becomes tolerable after the advent of human domestication, the provision of communal protection and safe shelters. The very fact that menstruation is still a taboo subject indicates that its role in human relationships is of great significance.

3:9 Facial expressions:

The Cambridge University specialist in human evolution, Professor Robert Foley, says (1987):

> "A further key characteristic of modern humans is the range of social relationships into which they can and do enter... it must be based on one attribute that undoubtedly has a biological and evolutionary element. This attribute is the

> ability to recognize and distinguish between large numbers of individuals, and to adjust behaviour accordingly. This is achieved by smiles, laughs and frowns that make the human face such an effective communicator."

In 1871, Charles Darwin wrote, "The human face is unique in its variation, its tremendous control of facial expressions, and its ability to transmit a wide range of messages."

Just why we need to 'transmit' such 'a wide range of messages' and 'distinguish between large numbers of individuals and adjust our behaviour accordingly,' to an extent far greater than our near relatives the great apes, has not been satisfactorily explained.

However, once viewed in the light of motor-control, *Homo sapiens* relationships do face many difficulties (see Chapter 6) and these must be negotiated and enacted through some method of signalling. Little wonder then that this complexity has led to a major enhancement in the range of facial expressions (and vocalizations) available to the hominins. Not only were these means of communication already in situ in a rudimentary form but once enhanced, they would be ideally suited to forge, enable, negotiate, consolidate or dissolve all the new patterns of complex human relationships.

The social signals expressed by the face make use of the mouth, lips, cheeks, facial skin, eyes, eyebrows, eyelids, forehead and ears to convey their meaning; they fall into seven categories (Tompkins and McCarter 1964):

1. Interest and attention
2. Surprise
3. Anger, rage
4. Fear, terror
5. Disgust, contempt
6. Sadness, anguish
7. Happiness, joy.

All are common to animals apart from disgust, contempt, and surprise, which are specific to humans. However, it is the use of, and variety within, each category that has undergone expansion, giving humans their wide repertoire of facial expressions. Similarly, the hands of the hominin, compared to the ape, have shown a significant increase in their use as a means of expression and communication. Approach-behaviour in an environment of motor-control is fraught with problems. Facial expressions that can range from a scowl to a smile are an instant indication of a likely response before any physical contact is made. As domestication and relative safety for the female increased, these facial signals gained a complexity which added to the meaning (or deceit) of language (see Chapter 9).

3:10 Voice box and speech:

Effective cooperative defences to motor-control necessitate a system whereby information about the identity, position, motivation and behaviour of other hominins can be speedily transmitted. Vocal communication is ideally suited for this purpose; however, living in a social group subject to the stresses of internal and external motor-control necessitated a major expansion of the original system of primate calls (see, Heteronomy and Language Chapter 9). All that is required here is to note that the anatomical changes in the human larynx are likely to have strong links, in the controlee, to the defences and distress caused by motor-control, and in the controller, to the efficient enforcement of motor-control.

3:11 Enlarged brain (encephalization):

Over the course of hominin evolution, the brain has enlarged and undergone profound changes from the original primate structure. All hominins have faced a constant threat of coercion from members of their own species. This resulted in a loss of autonomy and a rise of heteronomy not seen before in any animal species, and it gave rise to a top-down mode of mental operation (rather than the bottom-up mode common to all animals). The expansion of top-down heteronomous control also gave rise to a need for increased memory storage; hence the major changes in the neural networks of the enlarged hominin brain. These will be discussed more fully in Chapters 6, 7 and 8.

3:12 Hominin self-domestication or passivity:

Motor-control has led to the selection, in the controlee, of the ability to submit to and comply with the controller's directions. This acceptance of heteronomous control by the hominins has enhanced hominin survival but we have paid an extremely high price with the loss of our full autonomy.

This process of domestication required physiological changes: the fiery, animal defences that protect autonomy has been bred out of us and we have been forced to adopt a top-down 'responsibility or accountability' for all our actions (see Chapter 11). More than this we have become physically weaker than chimpanzees and this implies that a loss of strength has been selected, possibly because it reduced the injuries that stem from the bodily contacts and the coercion of motor-control.

Most feral animals are intolerant of motor-control, or even of having their flight distance encroached upon, and when they are captured they need to be sedated otherwise they react violently (become wild) and often die from panic (vagal inhibition). This would also have been the situation when the protohominins first experienced motor-control and the species has selected out the panic reaction and developed an acceptance of heteronomy and compliance in less than six million years. This unprecedented loss of autonomy has had serious consequences, and these are discussed in the following chapters.

Domesticated animals are more tolerant of motor-control than wild animals because they have been rigorously selected from the more placid strains. But this selection has meant that they have become extremely vulnerable and consequently they have suffered grievously under the excessive controls and punishments of humans.

Similarly, hominin evolution has also selected for an ability to withstand motor-control, which amounts to a self-domestication. Firstly, regarding maturation, the altricial period has been extended, increasing the age before which motor function and autonomy are fully established. Then the ability of unweaned young to suspend autonomy for short periods, when they are moved to safety by their dam, has been maintained throughout life and the ability to solicit motor-control to be carried has been expanded. All of which allows the young hominins time to develop before they are exposed to the full trauma of motor-control, and allows them to be carried, beyond weaning age, without it resulting in a serious loss of autonomy. Humans have also developed a prolonged period of adolescence (much longer than animals) in which the females are protected from coercive mating and both sexes have time to adjust to the pressures of heteronomy and learn to conform to the desired rules of adult behaviour before they become parents.

There has also been a selection in the male that has 'tamed' the urgent instant animal sexual response to the female. This domesticity has created a respite for the female especially in the family home, even though she is permanently receptive. In addition, there has been a selection for female passivity (see Chapter 8) that allows her to accommodate the continuous relationship demands that stem from permanent receptivity.

Clearly, all this has necessitated major anatomical and physiological change in the hominins. Genetic variation will mean that some individuals (controlees) are more able than others to tolerate unsolicited motor-control and the controllers will vary in their ability (or lack of empathy) to enforce these controls, behaviour not seen in the pre-motor-control world of animals.

I suggest that each of the above anatomical and physiological adaptations can be understood as resulting wholly, or in part, from the selection pressures of motor-control acting upon the hominin line. A list in summary may be useful:

1. Bipedalism
2. Reduction of hair cover
3. Reduction of canine teeth
4. Breasts on non-lactating females
5. Menstruation and engorgement of the womb
6. Facial expressions
7. Larynx (voice box) and speech
8. Encephalization
9. Hominin self-domestication.

Some of the behavioural and cultural changes that have occurred because of motor-control will be discussed in Chapter 4. The division between Chapters 3 and 4 lacks precision mainly because behavioural and cultural changes usually involve some element of anatomical and physiological change and vice versa. Nevertheless, those adaptations, such as domestication and the wearing of clothes, that are understood predominantly from the behavioural or cultural perspective are discussed in the next chapter.

CHAPTER FOUR

DEFENCES TO MOTOR-CONTROL, BEHAVIOURAL, SOCIAL AND CULTURAL ADAPTATIONS

This chapter focuses upon changes in hominin behaviour that are likely to have arisen as responses to the threat of motor-control. Behavioural changes such as food sharing, weapon use and the wearing of clothes are well known; what is new here is the need to consider them in relation to motor-control and the controlee's vulnerability to a loss of autonomy.

4:1 Selecting safer habitats (trees, caves, seashore):

Once the hominin female is unable to maintain her anoestrous state she becomes permanently receptive and so she needs to seek any site, habitat or natural features that will give her some protection from coercive males. Just as a prey animal tries to avoid its predator, so a female hominin, whenever possible, will try to avoid males from other groups who may be intent upon abduction and control. Avoiding hominins within one's own group who are also intent upon control is a much more complex problem that eventually gave rise to the establishment of a social contract and laws enforced by punishment.

At first, it would have been beneficial for the females, wherever possible, to remain close to clumps of trees or a cliff edge, where old safety patterns could be maintained (or re-established). Similarly, coastal living, by staying near a lake or seashore (Hardy 1960; Morgan 1982) the females could enter the water for protection (providing there are no aquatic predators). Mating in deep water would prove impossible without cooperation, and it may be the only natural habitat (other than a cliff edge) where the hominid females, once they had lost the ability to climb, could effectively establish and protect anoestrus.

Failing that, dispersal is an alternative strategy of selecting safer habitats; by moving ever onward out of Africa and then keeping ahead of other groups into lands originally free of hominins, our ancestors eventually walked to Europe, China, Australia and the Americas (and today we are planning to go to Mars). No other primate species has spread to every continent in this way and without any other satisfactory explanation the threat of motor-control needs to be considered as a likely explanation. Dispersal would have been a successful way of avoiding other hostile groups even though there is a cost to the group in staying on the move. Contact with others would be minimal and safety maintained (to the best of hominin ability) within the domesticated family group. Only when the population density gets moderately high would regular contact be made with other hostile groups, which would necessitate plans to defend home sites such as caves and camps with fortifications and weapons.

4:2 Food sharing / food denial:

> "Food sharing occurs in all human societies, and, given its relative rarity among non-human primates, is taken to be a key human characteristic."
>
> (Foley 1987).

Archaeological evidence has been found of the occurrence of food-sharing amongst the early hominins 2.5 million years ago. G. Isaac (1978) has suggested that food sharing is a consequence of meat-eating; males and females may forage separately but need to share the food when they return to the home base. This, in turn, would give rise to the need of maintaining and protecting a base camp that later would have the benefit of a fire for protection, warmth and cooking. It is suggested that the dynamic of this activity is premised upon male power and strength – "I'll get the meat," and "You get the herbs," but it is more likely to reflect the fact that after the advent of motor-control a lone female hunter out on the plains, away from home base, would herself be 'hunted' and coercively mated by the males of her own species.

Motor-control has implications for all aspects of hominin behaviour but particularly so for the daily necessity of feeding. In the animal world (apart from pack-hunting carnivores) the individual must find its own food input. Once the female is restricted in her foraging activities then she needs to be supported, to some extent, with food sharing arrangements. The use of hands and arms for carrying makes food provisioning possible but it is the threat of motor-control that turns it into necessity.

There is also another, perhaps more serious, complication and that is food denial. Deliberate food denial, either the physical removal of food, or the denial of access to food sources, is so simple to achieve with hands and motor-control that it, too, must have become common practice after the onset of terrestriality.

Food denial is very disruptive to social relationships and it creates a further problem, for as well as being able to withhold food, the controller is also able to gift or share food with those (and only those) he/she so chooses, perhaps in return for sexual favours or work.

Having entered social relations to this extent, food now becomes a reward and denial of food a punishment, and so it becomes a suitable means by which to manipulate and control others (as well as a channel to assist others – altruism). For adults, to be trapped in a situation where essential energy inputs can easily be denied by others is maladaptive, but this is just one facet of the vulnerability that is created as soon as hands are used to manipulate and motor-control others.

4:3 Home base, food carrying, pair bonds, defensive family groups:

The home base can provide protection from motor-control and new defences would be created whenever possible, for example, hiding in caves, using weapons such as sticks, stones, spears, fire brands and clubs, or creating taboos to restrict male encroachment.

The problem for the female is that these defences must not be totally effective; she cannot live and reproduce in a male-free environment. A compromise is struck in pair-bonding and polygyny: in return for permanent access for a single 'trusted' male, where the complexities of solicited motor-control can be explored, the female hopes to gain protection from unlimited males.

The division of labour comes about not because the female cannot hunt, but because she needs to stay near to home, for protection. The threat of capture, and rape, is a most powerful sanction against independent female hunters. Even at the home base, a single male cannot protect a female from a group of males so, for offspring survival, the males would have to group together (today with the aid of the law and police) and defend the extended family from unsolicited (and later, disapproved of) relationships.

Here is the basis of many of the social bonds, altruism and the concept of women as property, which are so characteristic of human behaviour. Without cooperation, assistance and defence of the group, every individual is vulnerable. The most powerful of these groupings is the multi-male group which the female relies on for protection from outside males (or from inside 'rogue' males). Extremely complicated social relationships result from these interactions which drive the need for increased brain capacity. In a world of motor-control those who were unable to maintain these complex cooperative defences would be at a serious disadvantage.

Carrying food to the home base mimics the feeding of altricial young. Hominins have a prolonged dependency (due to coercive motor-control) where food-carrying has become the norm. Mothers and their young are vulnerable to motor-control, rape and abduction, and so they are corralled in the safe home base. Once the females found a safe site, with naturally protective features, they would be reluctant to leave. Eventually this would lead to the depletion of local resources necessitating an increased expenditure of energy for food importation or the establishment of a new base in a more bountiful area.

For any domestication to have taken place, male animal sexuality must have undergone a further change. The natural, hormone-controlled sexuality of primates, which turned into the coercive sexuality of the early hominins, had now become overlaid with the inhibitions and repressions of domesticity (and later civilization); much of this change will have resulted from the new dynamics of the female's permanent receptivity. This remarkable phenomenon has influenced the male for it reduces the completion at the time of the female's ovulation. Indeed, the pair-bonded male is virtually guaranteed access to the female when she ovulates at any stage in her life. Therefore, the male's biological need for absolute 'urgency' of the sexual response can now be held in check or delayed without suffering the breeding exclusion that would otherwise have befallen the protohominin male, who had to compete and respond when the female was in oestrus, or miss a breeding opportunity for up to four or five years.

In conditions of domesticity, permanent receptivity creates a new problem for the female who now has a permanent obligation to mate. She cannot defend anoestrus, and so she is expected to mate at the behest of the male. This has advantages in that it is safer for the offspring and more survive but it is a heavy burden on the female.

The most effective defence for a female who is vulnerable to motor-control has come from living in a society that can maintain a social contract. When a female has protection from a strong close family, a supportive local neighbourhood, a police force, and a system of available justice, then she is relatively safe. But even within a protective family, father, uncle, brother or cousin can represent a concealed and serious threat.

Today, in the West, we have reached a situation where protection is mostly so effective that a female can (if she so chooses) provocatively advertise her sexuality and still say no to the male and, with the assistance of the law, be able to enforce that prohibition. In other situations, where the female has less support, she needs to be much more circumspect. However, effective protection can bring its own set of problems; it is also possible to find a situation where the family is so over-protective that the daughter is denied access to any male, let alone the male of her choice. This illustrates, succinctly, the human female's dilemma; defences that rely on others do not provide her with the absolute safety enjoyed by the autonomous, anoestrous female animal. The human female always needs to rely on others for assistance, but social defences can break down and when they do her underlying vulnerability returns.

Hominins can either be a force for the protection of autonomy or for overriding autonomy; violence and security are locked into a long-term struggle that is played out on the field of sexual relationships, where the female must either submit to a male (or males) in her group who provide her with protection or succumb to more aggressive males from other tribes.

There is a constant tension between these two forces and history shows that stable societies with strong defences always eventually collapse, for reasons such as food depletion, high maintenance costs or attacks from outside with new technologies or more focused aggression. This rise and fall in levels of safety is inherent in a species that cannot guarantee autonomy for any individual.

4:4 Agonistic buffering:

Another defence available to the female (closely linked to the phenomenon of maintaining breasts on non-lactating females, see Chapter 3) was to extend the practice of agonistic buffering. By holding her child (or a relative's child) close to her body, as if they were nursing, the female would, initially, be treated with much greater respect, for generally it is not in the male's interest to harm a mother with a child. However, the males, accustomed by now to coercing the females, would not continue to allow a child (particularly another male's offspring) to be used as a barrier to their demanding sexual desires so this defence has its limitations. But note that the use of agonistic buffering, and the wrenching away of that buffer, increases the incidence of unsolicited motor-control experienced by the child, compounding its problems.

Food provision ensures that the children stay in the home base near to their mothers, and other adults, whose presence will have some restraining effect on the marauding male. Hominin children are very vulnerable to abduction and they require parental protection until they are fully mature. Humans have a delayed maturity (compared to other primates) to give them a better chance of defending themselves, which makes the parental protection onerous and prolonged. Other primates can always run up a tree even when young, so their autonomy is not threatened in the same way as the hominins.

4:5 Building defensive strongholds and use of weapons:

Humans build defensive strongholds against members of their own species (even their own tribe or religion). These are not simply animal-type disputes over food or territory where intruders are driven off or killed. These disputes are about status and possessions based on an underlying fear of loss of autonomy and motor-control.

It is this anxiety that drives the frenetic elements of human behaviour because the tension of living with the possibility of bondage, rape, and slavery must be discharged and this creates many more difficulties for hominin society than death (natural or violent) does in the animal world. Death for the autonomous animal presents few problems for it is programmed (bottom-up) to deal with the complications that emerge at each stage of the decline. Whereas, being subject to motor-control creates such intractable problems that may never be resolved.

Gradually, the defence of the home base became more elaborate with fences, gates, stone walls, fortified hilltops and castles with facilities such as dungeons and prisons for the captured enemy. The area defended increased in size and it is now the violation of a country's borders that is an act of aggression that requires retaliation. All of this is so well known that it does not need any further expansion here, only to note that it all requires a remarkable expenditure of time and energy (that animals have no need to invest) and, to reiterate, without motor-control this level of intraspecific fear would not have materialised, indeed hatred can only emerge from heteronomy and motor-control.

Any loss of autonomy necessarily has serious repercussions, first, leading to the need to increase safety and protection (even pre-emptive strikes) and second to a prolonged search (religious and philosophical) for why we so readily become victims and controlees. All this thought and planning requires an increase in top-down brain power to maintain a necessary vigilance and defensive cooperation, none of which is needed by the animal because its autonomy is virtually never threatened by any member of its own species.

4:6 Clothes, nudity, modesty and shame:

Humans are the only animal to wear clothes; the accepted explanation is that they provide beneficial protection from the cold; this is technically correct in terms of heat loss, but it does not answer the deeper question of why we have deliberately chosen to make these man-made improvements to nature. This was the first time in over 200 million years of mammalian evolution that it had happened, so we are looking at (and trying to explain) something unique to the hominins, noting that several animal species (including those the hominins were hunting) had already evolved to cope with arctic conditions without any assistance other than a natural increase in fat, fur or feathers. Clothes may have speeded the process of the colonization of colder regions but there was no reason that the hominins uniquely needed this unnatural speed other than the angst that stemmed from their fear of conspecifics and the threat of motor-control that drove them to adopt new behaviour. This also needs to be seen against a background of hair loss, for the hominins have seen a reduction of body hair and presumably there must be some relationship between this fact and the wearing of clothes. Rather than allowing time for their hair to grow to give them added protection in colder regions, they have taken the reverse route, lost their hair and put on clothes (see Chapter 3). However, the additional point to stress here is that protection from the cold gained by wearing clothes is much the same as putting a horse blanket on a horse in winter, but the fact that the horse benefits from a protective cover does not mean that it then goes on to develop a need (like that of humans) to permanently cover its genitalia. It carries on as before (like all animals), unconcerned about its nudity.

So, there are two issues here connected to clothes. First, the unprecedented wearing of skins for warmth which probably took place as soon as the hominins moved into northern hunting grounds between one hundred and one hundred and fifty thousand years ago. Second, the heightened awareness of the genitalia that led to the perceived need for them to be permanently covered (even in warmer climates). This need is driven by an urge that is beautifully summed up in the phrase in Genesis, "Who told you that you were naked?" I suggest that this vulnerability arose out of heteronomous control and the resulting permanent receptivity, which inevitably led to the defensive position of shame and guilt. Both clothing-related issues illustrate the hominin discontent with the animal status quo. Any search for change and improvement, outside of the given bottom-up programmes, is a sign that the individual is troubled and restless and the reason for this unease needs to be sought rather than simply claiming it results from having a higher intelligence.

Unknown to the author of Genesis, early man, in the cold north, had taken to wearing clothes many years before the Middle Eastern tribes began to be worried about being naked. Indeed, females in equatorial Africa have only taken to wearing brassieres in the last hundred years encouraged by the Christian missionaries who thought their task was to 'civilize' the indigenous population. In addition to these regional variations there are religious variations where it has been decreed that the female must cover her entire body. Sexual guilt and stricter moral and dress codes are closely linked.

As housing became more common it could have been possible to wear clothes in the cold outside and take them off when inside the warm building, but this does not seem to have led to a return to nudity. Menstruation probably had an effect here for clothes help to conceal the menses and give the female some privacy. And it must be remembered that clothes cover and protect the female body, which is of some help to her when dealing with the coercive male.

Clothes protect the female from the male gaze and interest. Permanent receptivity creates the unease that females experience regarding their pudenda (defined as, 'that about which one should be ashamed'). As soon the males had a continuous interest in the permanently receptive hominin female genitalia then it was inevitable the female would be aware of that interest. This is not the case for the female animal for she is of little interest to the male throughout the whole of the time (most of her life) that she is anoestrus. Close attention from the male over the (short) period of oestrus is biologically essential for an animal species to breed but once the hominin female became permanently receptive then the attention becomes constant and this amounts to an intolerable invasion of her personal space, from which the wearing of clothes provides a small but welcome relief.

Clothes also provide the vulnerable female some protection in that they need to be removed, or rearranged, before the sexual act can take place. This creates a situation where the removal of clothes can be used as a bargaining chip by the female, for if she has full social protection she can then decide who can, and cannot, see her nude body. Today the female will rarely expose her body other than in the privacy of her bedroom and this means that a nude human female is a very rare sight. This is in stark contrast to the open nudity of all animals where every animal's body is seen by all in all circumstances, a situation that is integral to their natural grace. In the longer-term context, the hominin female has a need to have restored to her the autonomy that the anoestrous animal enjoys (that is, not to be permanently receptive) but until (if) that time arrives her ability to refuse or bestow favours can be a small but symbolic gain on her earlier plight. Nudity (which for animals is the open display of the truth of nature) has acquired a value which can be cashed by the human female for bonding or staving off the worst excesses of coercion. Further, if she has adequate protection, the female may use this new-found value (power) to exploit the males, by manipulating their sexual cravings to provide her with benefits, to the extent that female nudity now can become a favour, a gift or a transaction for profit.

Clothes create a major difficulty for male curiosity. The male is programmed to monitor the female genitalia for their oestrous condition; clothes block any everyday glances or inspections that animals consider normal and essential. Similarly, urination and defecation cannot take place without the removal of clothes and as the exposure of the genitals is now deemed to be so offensive or shameful these essential acts of waste elimination need to take place in private. The essential information (identification, hormone levels, status and levels of fitness) that animals gain from sight and odours of urine, faeces and genitals is not available to humans and they need to make up this deficit by other less efficient means – mostly by talking.

Step back and consider what this means; contrast the freedom of animals to expose their bodies and to urinate or defecate as and when their body dictates, with the nervous human need to find a private, appropriate place where one is not seen, ridiculed or tormented. In an overpopulated world sanitation obviously dictates some constraints but that should not necessitate an abandonment of the natural animal stance to these matters. Even breastfeeding, that purest and most essential of behaviours, is included in this list for it is deemed offensive to show a nipple in a public place.

The making of clothes is a dextrous process requiring two hands freed from other constraints. Like all other human artefacts, clothes are created out of the need to find better defences to motor-control. These defences have over time become nuanced and refined so that as well as having a practical function clothes are now able to signify fashion and status. The provocative element of fashion is of interest because it shows that many females now feel so secure that they can enhance their status by attracting and teasing the male who, because he is domesticated (has the new-found ability to restrain himself), he is now considered to be, in need of titillation.

Here is the dilemma that runs throughout the whole of this project: human relationships are so troubled and disturbed that we concoct reasons to conceal obvious biological truths. Without an understanding of motor-control, coercion, loss of anoestrus and permanent receptivity, it is not possible to come to an understanding of why clothes were invented, worn and turned into a profit-making industry.

Clothes are the perfect subject matter with which to start to explore the way in which humans first questioned and set down some of the early ideas about their origins and their behavioural predicaments.

Normally any reference to religious stories would have no place in a book claiming scientific credence. However, I take the view that the authors of some religious texts (particularly Genesis 3 and some Zoroastrian texts) were the 'scientists' of their day, trying to achieve a greater understanding of the human condition within the framework of their current beliefs. To think out the key concepts of the Genesis story while sitting in a desert village, over two and a half thousand years ago, pre-science, pre-Darwin, pre-evolutionary biology, is one of the most outstanding achievements of the human mind. If the present reader sets aside the idea of a personal God in a fictitious heaven (which has already lost virtually all its credence) and replaces it with the philosophical idea that the concept of 'god' has been created by mankind as the perfect foil with which to start to tease out the problematic essence of human nature, then we can further our understanding by heeding certain parts of the Genesis story.

Also, this is the literary tradition by which this subject has been maintained in the public consciousness for over 2,000 years (way before it was called anthropogenesis) and we need to add to that tradition from a secular vantage point. We need to find a way to be able to acknowledge the abundance, stability and variation of nature, which includes the fact that every individual is created and given by nature as an unsolicited gift without the parents having to do anything whatsoever, apart from play out their provided bottom-up mating and nurturing programmes.

There is also a need to be able to discuss how an animal comports itself in relation to not only the perfection of its birth but also to the fact that it comes into the world equipped with a full set of bottom-up programmes (which are autonomically updated with environmental information) that ensure, to the best available extent, its survival. With this inheritance it is then effortlessly able to 'bask within' or be 'carried upon' these programmes because they (without conscious effort) enable the individual to exist in the most efficient way within its species' environmental niche.

Following the realization (appreciation) of this awe-inspiring situation and understanding that this inheritance should apply equally to mankind and animals, it may be possible to connect this understanding with the term 'god' as a metaphor for these natural (rather than supernatural) facts. In our present distressed, distraught condition our scientific use of the term 'nature' falls short of an understanding that is sufficiently aware of this major property – that it is simply there (habitats are provided, given or gifted, in the non-personal sense of those words), as well as the fact that neither we, nor the animals, must do anything to ensure evolution's continuation (success) for it operates on a time scale beyond an individual's lifespan. (Indeed, without this deeper understanding we will continue, in our crazed search for safety, to damage and pollute. Global warming and overpopulation are already exacting an excessive pressure on the environment.)

Animals exhibit an indifference to all this, in the (so to speak) certain knowledge that their best interests have been accounted for within the bottom-up programmes that they inherit from their parents. Humans have a need to understand this animal ability to 'trust', without question, the overall operation of nature. They also have a need to understand that their life within nature's bounty should be lived in a similar way to that of the animals, expressing indifference, basking in the free ride that life has bestowed upon them (discussed again in Chapters 10 and 11). (It is a very minor point in the overall scheme of things if some wish to call this 'trust in God' rather than 'trust in nature', for ultimately it amounts to the same situation.)

In the Genesis story it is said that, as God drove Adam and Eve out of Eden, he fashioned animal skin tunics as a tangible reminder of their sin, transgression and knowledge of death, which they had brought upon themselves. Clearly, God saw nudity as natural, for when he first saw that Adam and Eve in the Garden were wearing their loincloths he cried out in despair, "Who told you that you were naked?" meaning that the natural, unashamed awareness of nudity (which is the animal experience throughout nature) is clearly the desired norm.

That is, we should be ashamed of wearing clothes, ashamed of our fallenness, ashamed of the fact that we see our natural nude body as a source of anxiety and shame. But under another remarkable inversion that humans tend to perpetrate, the reverse is now true: nakedness is a symbol of backwardness and savagery, animals are bestial in their nudity and humans are civilized, respectable and gain status by wearing their clothes. Natural (God-given) nudity is not just sinful, it is a punishable offence in law; nudity has become indecent and offensive.

Stephen Gough, who has become known as the Naked Rambler, has spent seven years in jail, much of that time in solitary confinement, for indecency, that is failing to wear clothes in public. At a recent trial in 2012, Gough said, "There is nothing about me as a human being that is indecent, or alarming, or offensive… So, this provocative nature that the man (Robertson, the prosecutor) is suggesting that I have is nothing of the kind. It is me, standing up for what I am… I have nothing to be ashamed about. I am just a bloke standing up for what I am."

In contrast, at the same time as Stephen Gough was in jail, others (mainly female) exhibited their naked bodies in clubs, pubs, on film, or the Internet, free to exploit the commercial value of their nudity.

Does a male otter say to his neighbour, "Put your clothes on before you venture out onto dry land"? Or, does a female otter say to her mate, "If you catch me five fish I will show you my pudenda"? Does a cow say to her calf as it seeks to suckle, "Not here, darling, the others think it's dirty"?

In October 2014, Stephen Gough's appeal to the European Court of Human Rights was dismissed and he remains in prison. If the animals of the world had heard this judgement they would have fallen about laughing. How could the judge (on behalf of all of us) be so out of tune with nature (God) and all the animals? What does it mean to say that nudity is indecent?

Effectively what the judge is saying, without being aware of the fact, is that the human female, having lost her ability to establish anoestrus, needs assistance to protect herself from the male and that the wearing of clothes has come to play a major role in that process. Stephen Gough is entirely innocent but the next woman he meets along the road has no way of knowing that, for he is a male and all human males have the potential to act coercively. Before Stephen Gough (or any other male) can regain his nudity, the human female will have to have her period of anoestrus fully restored and be confident (like all female animals) that she is safe from the (unclothed) male under all circumstances.

4:7 Individual psychology:

> "Some *Homo sapiens* exhibit behavioural patterns that are apparently neutral or maladaptive in terms of the inclusive fitness of the individual."
>
> (Hinde R.A. 1987).

For example:

1. Self-sacrifice
2. Gaining and maintaining status, wealth, and power not directly associated with reproductive success
3. Performing debilitating work
4. Gluttony
5. Anorexia
6. Obesity
7. Drug addiction
8. Sexual perversion
9. Celibacy
10. Religious ritual
11. Self-mutilation
12. Violence within the social group
13. Mental illness.

The logic and dynamic of much of this neurotic and psychotic behaviour stems from the experience of losing autonomy and having to find responses to exacting heteronomous controls that are set by controllers who cannot be trusted. Maladaptive behaviour can be understood as a necessary strategy for individuals who are attempting to defend (or draw attention to their failure to defend) their autonomy when it is seriously threatened by, or lost within, the dynamic of their interpersonal relationships (see Chapter 10).

4:8 Other complexities relating to motor-control:

The need for the elaborate human behavioural defences stems from the threat of a controlling attack from members of one's own species. Each attack and defence involves, in some form or other, punishment or reward, control or freedom, power or vulnerability, heteronomy or autonomy.

It is not possible to create a truly effective defence to our vulnerability to motor-control while we are ignorant of our predicament, for without a complete understanding of the difficulties we face, heteronomous controls will continue to override autonomy. Animals (apart from rare exceptions, see Chapter 2, Endnote 6) do not benefit or prosper from violating the autonomy of another member of their own species. Hominins have broken this golden rule and opened the floodgates of a new type of intraspecific predator/prey relationship where individuals or groups can prosper by overriding the autonomy of others.

Once collective security is established it tends to increase in size, the family becomes a tribe, the tribe becomes a territory, which becomes a kingdom or state, and the state becomes a superpower. As it enlarges it inevitably becomes distanced from the autonomous needs of the individuals that make up the group. Their present-day needs may not match the needs of the state and so compliance must be forced upon them, which then results in a further loss of autonomy.

Coercive power is always Janus-faced, for what is protection for some is a threat to others and here lies the inherent ambivalence of motor-control. The universal problem of establishing and maintaining individual rights will remain a problem until the animal model is adopted where autonomy is never compromised by heteronomous motor-control.

Having inadequate defences to motor-control leads to a universal quest for safety, a continual, persistent exploration of all facets of the environment and of relationships, to attempt to understand the inadequacies of one's own position and responses. Only a species that suffers motor-control would experience the feeling of having been thrown into the world inadequately equipped to deal with its everyday problems. It is this genuine feeling of deficiency and insecurity that drives the pursuit of knowledge, education, culture and religion. It is at heart a quest for safety and enlightenment and it stems from the loss of autonomy which can only be restored in a world without motor-control.

The anatomical, psychological, and behavioural adaptations (discussed in Chapters 3 and 4) amount to only a partial solution to the problem of motor-control. None, singly or combined, eliminates heteronomous motor-control, none re-establishes autonomy and restores anoestrus to the human female. None provides the means to stop the victim (female or male) being chased down, caught and controlled by their own species.

They all improve defences in various ways, some work for a time until the controller(s) find a countermeasure which in turn requires a new defence. It also should be remembered that effective defences for some often require the control of others and this twist has bedevilled all the so-called solutions that have emerged so far in human history, for none have been able to restore full autonomy to all.

In trying to understand the extraordinary vulnerability to motor-control that humans exhibit, we need to become much more aware of the underlying mental stance to life that animals exhibit and in so doing we would come to understand that, if motor-control was not the central fact of human existence, our minds would rest in a state of calmness and equanimity.

To recap, here is a list of hominin behaviour, cultural and social adaptations likely to be associated with motor-control:

1. Selection of safer habitats
2. Food sharing, food denial
3. Home base, pair bonds
4. Clothes, nudity, modesty
5. Individual psychology
6. Tool use, weapon use, use of fire.

Other areas in which motor-control has driven human behaviour (see Chapter 8 and 9):

7. Work, building, agriculture, animal domestication
8. Justice, culture, civilization, language, and religion
9. Controls, coercion, punishments, imprisonment, slavery, rape and torture
10. Weapons, warfare – defence and attack
11. Entertainment, alcohol and drug use, play, fashion, sport
12. Possessions, ornamentation, wealth and status
13. Education, art, literature, science, and modern communications.

This amounts to the usual list of human abilities and achievements which for many has always been evidence of human exceptionalism. The point to stress here is that they should be re-examined under a new light – seen in the context of a species that is struggling to deal with a catastrophic loss of autonomy and forced to live under the thrall of heteronomous controls. None are part of any animal behaviour, which means that if we were somehow able to restore autonomy and shed these behaviours we would behave as, be known as, and classed (once again) as – an animal.

I suggest that all the above hominin behavioural, social and cultural adaptations have been invented, adopted, remembered, organized and enacted with the aid of others (and despite others) because of motor-control, coercion and heteronomy. This requires an enlarged mental capacity compared to that needed by the animal, hence encephalization and the top-down organization of the hominin brain. Motor-control creates impossible difficulties for autonomous governance and this will be examined in the following chapters.

CHAPTER FIVE

PHYSICAL RESTRICTIONS OF BEHAVIOUR

Before autonomy and heteronomy are discussed in greater detail this chapter sets out some of the ways in which animals can have restrictions placed upon their movements.

5:1 Restriction and Avoidance:

An animal may have to modify its intended movements and it does this simply by a change of plan or by avoidance. Natural features such as mountains, rivers, thick vegetation and climatic conditions such as drought, flood or snow may, from time to time, restrict an individual's intended movement but after a quick reappraisal of the difficulties and dangers involved, the intended movement is abandoned and readily replaced by another.

When an animal comes to a deep river, a tree or cliff, its behaviour is curtailed but this is not 'control' because there is no intention on the part of, say, a river to stop the animal accessing the land on the other side. The animal needs to vary its intended movements, which may present difficulties, but this does not amount to a serious biological problem; it is not a direct threat to autonomy because the individual is free to move in other directions.

Loss of habitat due to, say, climate change, drought or flood may be more restricting but even when starvation is the outcome the individual is free to move and make his/her own decisions, seeking food or sanctuary up until the moment of death. This is also the case with the life-threatening presence of predators, dangerous snakes, poisonous insects and large mammals that restrict the movements of other animals as they exhibit extreme caution. The individual adjusts to these restrictions and restraints and continues the best it can; sometimes the individual loses its life, as for example in predation, but none present the degree of mental complexity that arises from motor-control whereby a controller tells a controlee what, or what not, to do.

Dominant animals restrict the movements of intraspecifics and their 'control' has an intentional element; for example, alpha males will influence the movements of other males by denying them access to the females or the latest kill. But even here the subordinate does not lose his autonomy because it is left with a choice to back off; it simply needs to give up on his (never guaranteed) mating intentions, or plans to feed at the kill, for fear of sustaining unacceptable levels of damage in a fight for dominance.

The female's situation is more complex; for example, she may have her movements temporarily restricted by being herded by the dominant male. In this case, she must be in oestrus, or approaching oestrus, and within the breeding season, for the male to have any influence over her movements (remembering that it is also in her own mating interests to participate). Her personal space is not closed so tightly that she loses autonomy but on occasions it can become very restricted and freedom of mate choice may well be denied to her. Usually it is in her own interests to mate with the dominant male but rarely she may exhibit another preference;

however, this is an area that is notoriously difficult to ascertain with any certainty. When loss of mate choice does occur, it is for a short period, perhaps two days once or twice a year, while the female is in oestrus. Although this is a serious restriction it does not cross the threshold of a major loss of autonomy that undoubtedly occurs when the human female is held and coercively mated (raped).

Dominant animals can make subordinates adopt avoidance behaviour simply with an aggressive look, which conveys the threat of punishment. And by fighting, they can bring a subordinate to the point of submission, but the dominant animal does not have the ability to completely close a subordinate's space. The dominant animal can only be in one place at a time so there are plenty of other spaces for the subordinate to move into and it is not seriously damaging for the individual to have to move off to place X rather than place Y, whereas with motor-control, the hominin controlee, having lost autonomy, cannot avoid being held in place Z: she/he has no other option.

Avoidance of the other guarantees the safety upon which animal intraspecific behaviour is structured. When one animal deliberately challenges another, then a fight ensues which ends in submissive behaviour where the loser has either to back off and (say) stop pursuing the female, or move away from the food cache, or submit by exposing the vulnerable underbelly, which in effect says, "I won't do it again".

Subordination means that the dangers posed by dominant peers must be incorporated into the subordinates' safety/danger scales. Off a level playing field, they would want, for example, to eat the meat or mate with the female but are unable to do so because the alpha male does not allow it. Their behaviour is modified by backing down. This is normal animal behaviour, it is not absolute freedom of movement whenever, and wherever, the subordinate animal wants but it is always able to move somewhere else and so autonomy is never lost.

All these restrictions are readily accommodated by the animal because each individual has inbuilt bottom-up programmes that have evolved to deal with the constraints that the species may face.

Motor-control has created a new danger for the hominin (as well as those animal's humans domesticate) for it is the act of being held that ultimately enables the controller access to the controlee's mind and to insert any heteronomous controls he/she so chooses. Wild animals must change behaviour when threatened by dominance but even the largest alpha male cannot impose specific directions, such as, "Do this," or "Do that," or "Climb that tree and bring me ten figs," or "Catch me an antelope," or "You wait here," or "Hide behind that tree," or "Go there," or "Do this," or "Come here," or "Sit down."

Wild animals cannot be forced to comply with any directions.

The dominant animal does not have the ability to turn a subordinate into a controlee, that is, it cannot make a subordinate do anything other than move out of the way. This is because the dominant animal cannot hold the controlee past the point of submission (in fact or in fear of the fact) and this is a crucial difference between animals and humans.

The dominant animal does have the power to stop some subordinate activity such as mating or feeding, and it can make a subordinate run away for fear of an attack, but it cannot make a subordinate do anything else; this is an important limitation as it preserves the autonomy of the subordinate.

Whereas a human controller can (with the use of hands and the fear of motor-control) make the controlee comply with virtually any wish, desire or instruction that the controller chooses. An anoestrous female can be forced to mate, males and females can be coerced to work, confined in one place or banished, or made to kill others (and the controller does not necessarily have to be present). No animal has this power over its conspecifics; the contrast is stark. It is the essence of being human: having the ability to be able to hold another's body, get inside their mind and control their behaviour, and it has enabled the controller to make a controlee carry out virtually any action the controller chooses.

The dominant animal (if present) can stop the subordinate eating a piece of meat, or stop it mating with a chosen female, or stop it claiming a specific territory, but it cannot deny the subordinate all meat, mates or territory, nor can it make the subordinate eat a particular piece of meat, mate with a particular female or live in a particular territory, whereas the human can make another do all of these things and more.

5:2 On being seized, avoidance denied:

It is the subordinate's ability to avoid confrontation simply by moving away that limits the power and influence of a dominant animal.

Animals can rarely completely close the space of a conspecific; when that does happen both parties are programmed (bottom-up) to act accordingly, the weaker one submits, the dominant party backs off, that is, it stops fighting and does no further damage to the victim. The submissive animal registers the dominance and acts appropriately by contritely slinking away, thereby regaining his/her personal space which, under the rules of dominance, is slightly more limited than before but this is a small price to pay for remaining physically unharmed. If the dominant animal holds the subordinate past the point of submission, it could with claws, hooves, teeth, or horns, if it so chose, kick, bite, or claw the victim, causing serious injury or death, and this would be much like the actions of a predator. But a powerful inhibitor has been selected in the operation of intraspecific behaviour that stops the attack progressing this far. The dominant animal does not go past the point of submission to inflict serious damage and the chastened submitter can retreat unharmed.

However, with motor-control, a situation has arisen within nature where the dominant hominin can easily go past the point of submission without it resulting in serious physical injuries or death of the submitter. Life for the animal that loses (past the point where it submits) is not significantly different to that which went on before it was forced to submit. After the act of submission, the dominant animal does marginally increase its ability to influence the behaviour of the submitter, but only in the sense that the submitter becomes more cautious and circumspect and keeps a safer distance.

There is a major difference between making an animal change its original intentions by means of threats and bites, and making a hominin, under the coercive threat of restraint, act out behaviour that has been sourced in the mind of another and would be rigorously avoided by the controlee if she/he were free.

The simple act of motor-control means that avoidance has been denied to the controlee and a completely new world of heteronomous compliance is forced open. This is the 'new world' in which humans now live and it lies beyond the point where autonomy is lost. It is a world unique to the hominins and it requires different intraspecific behaviour for the individual to survive. To facilitate this major change a 'stop switch' in the animal brain has been overridden (selected out) that before would always engage to bring about a cessation of an intraspecific attack as soon as submission was signalled. For the controller to impose heteronomous controls he/she must continue past the point of submission and impose his/her wishes, desires and controls upon the controlee.

Many controllers are oblivious (deaf, blind) to the submissive gestures (cries, pain, tears) of the controlee when she/he is grasped and held, and they will press on with their assault, rape or enslavement without any instinctual inhibition. Hominins no longer back off the moment autonomy is overridden; we have lost our animal sensitivity to (empathy with) the plight of our conspecifics.

Once a hominin is held, avoidance becomes impossible and the only action open to them is compliance – to agree to enact the wishes of the controller and comply with his/her heteronomous directions. That is, to endure and enact the movements and behaviour that are required for rape, slavery, torture, imprisonment, servitude, work, compulsory education or any other heteronomous control to take place.

This takeover by heteronomous (as opposed to autonomous) governance cannot be achieved without a catastrophic increase in the vulnerability of the controlee who had to develop a new-found top-down mental ability to absorb and enact the controller's instructions. Behaviour forced upon another because he/she is denied avoidance is always behaviour that is enacted in a space that is 'past the point of submission', 'beyond avoidance', 'against the will'. This is a new territory within nature that has been opened solely by the hominins with their use of hands to motor-control; it has resulted in severe restriction to autonomous freedom and given rise to the imposition of heteronomous controls.

The next chapter outlines the ways in which the human mind accommodates these new heteronomous controls by adopting a top-down mode of organization.

CHAPTER SIX

BOTTOM-UP AND TOP-DOWN CONTROL OF THE HOMININ MIND

The first act of motor-control took place against a background of normal autonomous animal existence. The first coercive matings will have occurred spontaneously, driven by the natural, normal, bottom-up programmes of the male seeking mating opportunities whenever and wherever he could – which means that the newly found ability to motor-control was (for the male) no more than an unexpected fortuitous benefit of terrestriality.

This created an exceptionally difficult experience for the female. A dark period ensued where defenceless females would have been relentlessly pursued and coercively mated. Eventually, coercion became commonplace, planned, expected, experienced, witnessed and defended against. This would have been the start of a period of intense selection acting directly upon breeding females, where any small advantage that aided survival would have been passed on to their offspring.

6:1 How motor-control changed the mind's organization of behaviour:

I suggest that these pressures gave rise to the selection of the enlarged brain and top-down thinking, which enabled new mental strategies to be developed that afforded protection from the worst excesses of violation. The first strategy was compliance; the controlee complied with the controller's instructions and enacted his directions in the hope of reducing injury and punishment. Today, compliance is a mundane event but at the time of the first hominins it was an extraordinary event for it meant that the controlee enacted movements that were sourced in the mind of the controller. After finding ways to respond to heteronomous controls they would have learnt to organize collective defences to those controls and to impose, where possible, their own heteronomous controls upon others. Animals do not experience heteronomous control, so they have not developed the neural circuits needed to deal with these complex and unpredictable relationships.

To be able to easily differentiate between these processes I have called the way in which the human mind organizes its responses to heteronomous controls 'top-down' and the way that the animal brain organizes its autonomous behaviour 'bottom-up'. (1)

All bottom-up programmes function around the default position of avoidance. If difficulties arise for the individual animal, it abandons its behaviour and flees. The use of the hominin hands to hold and control others has blocked this strategy which meant that the bottom-up programmes could not be relied upon as a guarantee of safety. Being held immobilizes the controlee and on fear of further punishment the controlee must absorb the directions of the controller (this is a situation that simply does not materialize in the animal mind because the animal is always able to take avoidance action). Hence, the top-down neural circuits developed so that the controlee could comply with the controller's instructions. Eventually, self was able to monitor itself and reflect upon, rerun and juxtapose memories of the controller's demands, and self's own reactions to those demands, with a view to discovering ways to improve, or resolve, the difficult predicament in which it found itself.

6:2 The creation of top-down neural pathways:

Animals do not have the ability to catch, hold and control intraspecifics and deny them the safety that comes from avoidance (see exceptions, Chapter 2, Endnotes 2 and 3). This means that no animal can hold another in such a way as to gain direct access into its mind to 'tell' it what to do, or what not to do. An animal can influence another animal's behaviour by means of threats, submission or friendly gestures but no individual of any animal species (apart from the hominins) can manipulate another individual to the extent that it can implant instructions in its mind and force it to comply with exacting heteronomous controls.

It is the lack of a direct top-down mental connection between the brain of one animal and that of another that acts as a safety barrier denying entry to all inroads that may result in heteronomous control. In this respect, the protohominins were like all other animals and it was not possible for heteronomous controls to gain access into their autonomous system of governance. That is until hands were released from the constraints of the trees and an entirely different set of selection pressures came into existence.

The selection of top-down neural pathways commenced from the moment the first hominin female experienced the effects of coercive mating, it being impossible for the bottom-up programmes to organize defences that provided any meaningful form of protection. With no means of escape (because she is held) she has, somehow, to direct her own body to make the movements that the controller imposes upon her and if she is coercively mated she must act against the (true) information coming from her own body's hormones – that she is definitely anoestrus. She acts against what would normally be her own best interests and makes her genitalia accessible to the male and she does so in fear of the pain and injuries of the punishment that would follow if she did not cooperate.

This leads to the bizarre situation where she is forced to interpret (pre-speech) the controller's body movements, understand and comply with his gestures and directions, that is, she must marshal and organize her own body movements according to the wishes, desires, instructions of another (the male) rather than act in a way that her own bodily hormones would have freely dictated.

The neural pathways that process the information from this complex interaction have developed top-down because:

1. This is the way the controller's directions enter the controlee's mind (top-down, or outside-in).
2. There are no stored or inherited (bottom-up) programmes to deal with this evolutionarily new, complex, contradictory and unpredictable situation.
3. Only top-down mental organization is flexible and agile enough to comply with the range, variety, originality and unpredictability of the controls, directions, wishes and instructions that the controller(s) may enforce, given the short time in which the controlee must comply. These controls, sourced in the mind of another, are, at first, essentially unknown to the controlee (and sometimes even to the controller who may occasionally act upon a whim), yet the controlee has, somehow, to process all this information to be able to enact the instructions to the satisfaction of the controlee.

The original controller(s) had a simple (but threatening) sexual motivation and their initial aims, were focused upon assaulting the female. However, this (narrow) threat gave rise to the development of the top-down system of neural pathways that enabled the controller's directions to penetrate the controlee's mind and this structure was then available for the controller to pursue a much wider range of controls.

This meant that the controlee had to learn how to marshal, direct and enact virtually any heteronomous control in compliance with the controller's wishes. These new top-down neural pathways that have facilitated this complex interaction between the controller and the controlee I have termed the 'template of control and compliance'.

The template of control and compliance has developed out of the area of the animal mind that monitors dominance and deference. All social animals must be prepared to adjust behaviour depending upon their assessment of a conspecific's size, strength, age, health and status. These adjustments determine the dominance/deference behaviour within the group. Every individual has the ability, depending on the circumstances, to either act dominantly or deferentially. This ability to switch back and forth between opposite behavioural stances, depending on the relative power or weakness of the conspecific, is the underlying basis of operation for the template of control and compliance. In the animal this is done bottom-up but in the hominin its use has been expanded (because of motor-control) and it has come under top-down control.

Once the controller can get inside the mind of the controlee to direct heteronomous behaviour, then an unprecedented situation is created which means (meant) that the controlee not only had to understand, comprehend, (hear) the controller's instructions but also had to be able to direct her/his own body movements (top-down) to enact the behaviour desired by the controller – that is, to comply with the directions (movements) sourced in the mind of another.

At different times (in varying situations) we are all capable of being controlled (becoming the controlee) and we are all capable of imposing controls (being the controller) hence every individual has within them the neural circuits necessary to act out these two different roles.

For the controlee the section of the template dealing with compliance must have the ability to:

1. have a constant alertness to evaluate the power and strength of all conspecifics and the degree of danger they represent including the punishment they may inflict;
2. be receptive to the controller's trespassing instructions as they are forced into the controlee's mind;
3. be able to understand these instructions and convert them into the actions necessary for compliance (such as, sexual activity while anoestrous, work, or tolerance to restrictions of movement). Being able to understand instructions requires a wider responsiveness to sounds, such as shouts, grunts, growls, screams, cries and eventually words;
4. be able to balance (stack) the demands (sometimes conflicting) of more than one (sometimes several) controllers so that those with the harshest and most imminent punishments are granted priority.

For the controller, the obverse side of the template comes into play to facilitate the imposition of controls, so it can:

1. evaluate the power and strength of conspecifics with a view to dominance and control;
2. organize, marshal and expand the field of dominance into the ability to motor-control which enables the controller to use the controlee for his/her own advantage;
3. learn that having trespassed upon the female sexually then it was (is) also possible to trespass upon both male and female

> in many other areas, such as restriction of movement, or the imposition of work, in fact given sufficient power, compliance can be demanded in any area of social behaviour that the controller so wishes. Hominin trespass of the controlee requires the controller to show considerably less empathy towards the conspecific controlee than does animal dominance which does not seek (even temporarily) to take over top-down governance of the conspecific.

Compliance is ultimately motivated by fear of punishment, while controls and trespass were originally always motivated by gain, advantage and greed. As things have developed this division is no longer as clear-cut, since defences to being controlled have understandably taken the form of trying to control the controllers, thus the imposition of controls (in defence of the controlee) can now be motivated by concern.

The fight for governance between the controller and the controlee takes place within the battlefield of the template of compliance and control. The template holds the details of the compliance instructions (memories of how, when, where and the costs involved), which enables the controlee to enact the control at short notice. Within this memory bank is the facility to update the instructions to accommodate changes to the instructions or, if possible, make the compliance less stressful.

This ability to modify the enaction of the control may, in certain circumstances, allow a defence strategy to develop, seeking a better, or less arduous ways of proceeding, such as, "How can I modify these controls?" or even, "How can I get out of this situation?" Compliance, although it represents the safer route in a dire situation, still exacts a toll and always remains a danger to what should have been the autonomous governance of the controlee. So, the template of compliance not only facilitates the mechanics of compliance, but it also facilitates the search for better outcomes;

the two are inextricable. Thus, the controller and the controlee are locked in battle and the arms race between autonomy and heteronomy escalates.

To be able to avoid punishment, top-down compliance must be enacted the moment the controller desires. To be able to do this, compliance must be able to override any other response that may have been set in motion by the controlee's bottom-up programmes. This granting of priority (necessary in the face of physical punishment) has had dire consequences, for it facilitates heteronomous, rather than autonomous, governance. When two individuals are locked into a relationship of control and compliance the communication between them must be fast and clear and it is here that memories and auditory responses (eventually speech) play a pivotal role, for words are the foot-soldiers of heteronomous control (discussed more fully in Chapter 9).

6:3 How top-down mental organization developed complexity:

In the final analysis, it is the threat of being held and subjected to motor-control, together with the fear of subsequent injury or punishment, that forces the controlee into compliance. When examined closely, compliance with the controller's directions is seen to be a truly remarkable ability (vulnerability) but it also gave rise to a further extraordinary development, for once the controlee can operate under this new system of heteronomous governance it creates an opportunity (opening) for self to control itself top-down – that is, for self to override its own bottom-up programmes.

This has created a tripartite system of governance within the individual hominin mind, where governance of behaviour may originate from one or other, or a combination of these sources:

1. External top-down heteronomous controls and directions imposed by the controller. Originally, these controls were detrimental, but some are motivated by a genuine concern for the controlee and some of them may now be understood as being of benefit given that the controlee is struggling to live in a hostile world of motor-control without adequate defences.
2. Internal top-down control of self, by self (willpower) that has the objective of creating better defences to motor-control. This may be understood as a compromised autonomy that attempts to implement auto-heteronomous behaviour (which is a mix of heteronomy and self's adjustments to those controls or self's attempts to influence or alter those controls). For the human, what is left of autonomy is a compromised restricted form that maintains vital, or urgent, basic body necessities such as breathing, urinating, defecating, vomiting, yawning, sneezing and childbirth but even here these responses are subject to top-down restraints to ensure that they are performed 'correctly or discreetly'.
3. Pure autonomous bottom-up governance (as in the animal) is now unavailable to the human because of the restraints imposed by external and internal heteronomous controls.

These different sources of governance give rise to a complex system of intentions and it is often difficult (sometimes impossible) to be clear as to where a segment of behaviour was sourced or originated (whose idea it was in the first place and in whose interest it was carried out) and whether it is heteronomous, auto-heteronomous or autonomous behaviour, or a mix of the three?

Human behaviour is mainly auto-heteronomous behaviour, in which the percentages of the auto and the hetero elements vary at different times. To start to unpick all of this we need to examine top-down behaviour in more detail.

Complying with top-down instructions is an evolutionary

new development requiring new or improved neural pathways. Those who 'agree' to carry out the controller's instructions (that is, gesture that they will comply) have a major selective advantage, which would eventually lead to the selection of the larger hominin brain. If the female submits before she is beaten she is spared serious physical injury. If the slave carries his load before he is flogged, he is more likely to survive. If he does not understand what to do or why he should do it (why carry another man's load?) or, if he still refuses to do it after being beaten, then he is likely to be beaten again and more seriously injured. Only those who are alert enough to be able to think top-down and comply with the controller's instructions (that is, become a puppet) will be able to make this new (treacherous) accommodation with heteronomous control.

Self must enact the directions of the controller even though these directions are sourced externally. Self has somehow to convert a mental 'understanding' of another person's desires, wishes or instructions into her/his own body movements and enact them to the satisfaction of the controller. At this point an extraordinary situation has developed within nature that has done incalculable damage to social harmony, for wholly autonomous governance has had to accommodate heteronomous controls that force the individual to act in contradiction to all its underlying principles.

Most autonomous behaviour has a lead-in time that pre-prepares the individual before the activity takes place – hunger precedes eating, tension precedes a fight, tiredness precedes sleep. Whereas heteronomous controls are 'unpredictable' in the sense that the lead-in time is in the mind of the controller, such as, "I'm going to make her/him do it now," and this information is unavailable to the inexperienced controlee until the act is forced upon her/him. Without a lead-in time the controlee must be able to quickly catch onto, or understand the controller's intentions if she/he is to avoid any punishment. After time, experience would help the controlee to predict or anticipate the control of specific controllers, all of which require a major increase (encephalization) in mental capacity and memory storage. Compliance necessitates meta-cognition for there are now two or more perspectives (the controlee's and that of the controller or controllers) from which any behaviour must be considered before it is enacted. This has resulted in the much-vaunted increase in human intelligence but note, it has come into being for no other reason than that we have fallen victim to the heteronomous controls imposed by our fellow human beings.

From the external controls of the controller to the internal controls of the controlee, the template of compliance was the key to all subsequent hominin mental development, for its top-down route was (eventually) used to direct all forms of behaviour (not just sexual behaviour).

But its most startling, innovative and evolutionary new application was that it allowed self to turn in upon itself and, top-down, create and direct her/his own thoughts, thereby gaining the ability to override most of self's own bottom-up programmes.

Once self was able to direct its own thoughts, wishes and desires along the top-down neural pathways it meant that for the first time self was able to make itself comply with its own self-constructed top-down thoughts and plans – that is, self was able to create its own novel behaviour. If others can make you comply with heteronomous instructions (tell you what to do) then the compliance template that organizes and implements those instructions can also be used by self (fearing motor-control) to tell itself what to do. I suggest this is a decisive moment in the evolution of the human mind for the animal mind cannot get 'outside' of itself in this way.

Here is the birth of self-consciousness for it is inevitable that if self is forced to modify its own (first-choice) bottom-up behaviour to comply with the wishes of the controller then it will become aware of itself doing so. That is, self will become conscious of its own inability to defend autonomy and then become conscious of its own ability to construct compliant behaviour (and with that an associated anxious uncertainty as to whether that compliance will satisfy the controller).

And here is the underlying drawback of this new-found ability: it is formed out of the problems and difficulties that result from motor-control and this gives the top-down monitoring human mind a troubled, uneasy, self-critical, anxious focus on the world that is not experienced by the animals. They are blessed by having a bottom-up mind that has an enviable, equitable, accepting outlook on the world.

Thus, animals are not self-conscious in the same way as humans, they are not compliant to top-down controls and so they confidently rely on and function within their own bottom-up programmes. For humans, although the ultimate seat of governance remains with the bottom-up safety/danger scales, the effective seat of governance now resides in the top-down compliance template which operates the safety/danger scale specifically concerned with motor-control, outlined earlier, (Chapter 2, Endnote 4). It is impossible to overestimate the significance of this change, and the creation of this new seat of governance within the individual, for it has facilitated all heteronomous top-down social behaviour.

In addition, there are top-down controls where the relationship between the controller and the controlee is made more complicated by of the involvement of third parties. Broadly, these take the form of either, (a) the use of top-down heteronomous control of others to defend oneself from motor-control, (b) cooperative top-down control of others so that they can defend themselves from motor-control or, (c) top-down control of others so they learn to defend others from motor-control and other dangers. We will now look at these in a little more detail.

(a) Once an individual can create and implement top-down plans for him/herself it is a short step to go on and make plans for others so that they assist in these defensive endeavours, for example, "Don't go, if we stay here together he won't attack me," or "Run to get help." The controlee may also benefit from spontaneous assistance from others, who having gained top-down control of themselves may collectively try to attack the attacker; indeed, human empathy (beyond the mother–offspring bond) probably developed here as others come to the assistance of the controlee.

As soon as the controlee can gain assistance, a second tier of defence is created making it somewhat more difficult for a potential controller(s) to impose their controls. This means there is now coercive mating and defences to coercive mating and these may involve secondary coercion; he controls her, and she controls others to help defend herself from him. (Similarly, the controller may seek assistance to enforce his controls and the arms race escalates). To be able to persuade others to comply with directions and regularly offer others assistance is a remarkable new situation in evolutionary terms. Benevolent control, however, tends to become very complicated and the rules, regulations, contracts and restrictions that are needed to contain excesses and establish fairness can easily become oppressive. This is where society tries to carve a balanced approach, but it can never overcome the objection that heteronomous controls, whatever their purpose (aggressive or benevolent), are always the antithesis of autonomy.

The controlee has found a way to control not only others that are smaller or weaker but also those who are larger but willing to be co-opted into a plan, because, top-down, they have realized the benefits of cooperative defences, or attacks. These controls may also be covert controls based on some form of bargaining, "I will stay here and protect you and in return you will mate with me." Patriarchy, pair-bonding and marriage can be traced back to this moment; the female agrees because it reduces her vulnerability, but it is certainly not a return to autonomy.

(b) Here the mutual complexity of cooperative control increases. Those who have experienced (or understand) the operation of heteronomous controls can themselves benevolently control, assist and teach others to defend themselves. Plans and more elaborate strategies can be introduced that assist or direct the behaviour of others so that they construct their own defences. "Don't forget to run away when he comes," "Stay here and hide with us," "Try to keep together," "Don't go out there," or "Store this food so you can hide."

A female with a history of being abused, or knowledge of the abuse of friends, neighbours or relatives, comports herself in a different way to her more fortunate sisters. Her experiences will inevitably increase her own tensions and anxieties, and this will be apparent to her children and other members of her family. All of which will stimulate a desire in those who have sufficient empathy to cooperate to create better defences that in the longer term improve society. Throughout the whole of human history, it is this concern that has given rise to the quest for justice – a quest that is absent from all animal motivations because they are not motor-controlled, and they do not survive if their autonomy is violated in other ways (predation).

(c) Here the complexity increases yet again by creating another tier of defence. In the 'civilized' world, specialists are now trained to specifically defend others from coercion. This is where the best intentions of the military, the police, the legal system and the public service are displayed. Also, compassion has spread beyond concerns about overt coercion into other areas such as injuries and ill-health which require nurses, doctors and other health workers.

There is always an underlying problem with this level of interconnectedness and dependency: it relies heavily upon heteronomous instructions and controls which, even for those motivated by the best intentions, are still burdensome. Animals do not have policemen, judges, lecturers, doctors, nurses, veterinary surgeons; they do not need to resort to this level of complexity in their interactions because they are always able to look after their own needs. In avoiding all forms of heteronomous control, they are able, without top-down thought or planning, to defend their autonomy – autonomously.

As the complexity of the top-down defences grew, and involved more participants, it necessitated the metacognition of second-, third-, fourth- and fifth-order thinking – "I ask you, to ask him, to tell his mother, that she must tell her son that his daughter has to tell her sister to take care and protect her children from that evil man…" It is the urgency of the need to do something about the threat of motor-control that initially glues these levels together and links each order of thought with the one above. Without this dire necessity the original pattern (now deployed in situations other than motor-control, for example games of poker) of nth order thinking would not have needed to become established, for metacognition is present in the animal mind only to a minimal extent.

6:4 Comparison of top-down and bottom-up neural pathways:

The autonomous animal does not need top-down neural pathways for its mental activity, nor does it seek them. Bottom-up was the only mode of mental organization available throughout the entire evolution of motile life forms up until the hominin terrestrial transition, which means that the self-conscious top-down mental activity was unknown for the first ninety-nine per cent of evolutionary time.

The responses of bottom-up behaviour are selected by evolution. The responses of top-down behaviour are determined (created) by heteronomous control. Bottom-up flows effortlessly while top-down is contrived. Top-down cannot find equanimity because it is created as a response to the dangers of motor-control and a loss of autonomy. Bottom-up is the result of natural survival, in the whole of pre-hominin evolution nothing is manipulated top-down, and it is not subject to anyone's desires or controls. The bottom-up world is the world of nature; animals live in this world because they do not motor-control, whereas humans live outside of nature in a world of top-down mental control because they are subject to, and they subject others to, heteronomous motor-control.

The bottom-up programmes freely select and enact responses, one after another, according to their inherited order of priority. This is a harmonious process without friction, questions or problems. The responses are enacted without internal dispute because the priority response is the one best suited for survival at that moment. Responses have no interest in being selected, nor are they disappointed by not being selected, the responses are simply available to be selected in the appropriate situations determined by the safety/danger scales.

Not until motor-control made it impossible for the individual to live autonomously was top-down control able to gain a hold. Once the flow of autonomous governance is disrupted – the sweet reliance of 'being taken care of' (looked after) by one's own bottom-up programmes is lost; the individual must anxiously cobble together, as best he/she can, some sort of response to be able to enact the movements that the controller has deemed to be 'wanted' or 'correct' (that is, those that please the controller). For the controller to be able to place instructions and directions into the controlee's mind, which the controlee must enact, is a quite extraordinary evolutionary development.

An animal with a bottom-up mind does not question why this (rather than another) unit of behaviour has come up; it simply accepts it and enacts it. It has been delivered to its mind and body by nature through inherited programmes developed over years of evolution, with the implicit understanding that this action is the best for survival at this moment of time. Bottom-up is not charged with any responsibility to work out what to do next, in the next hour, or next week, or next year, because all necessary programmes are already stored in the safety/danger scales waiting to be triggered when conditions dictate.

In contrast, top-down organization needs to be aware of all its options because there are so many personal variations depending upon the wishes, desires, whims, threats and dangers that arise from having many potential controllers, and so it accumulates a cache of plans and responses that can be brought into operation whenever necessary. (See Chapter 11, Personality). Top-down mental organization is very different to bottom-up organization; it is a new mode of operation for the brain, it represents a major evolutionary change and it gave rise to the cognitive revolution.

The protohominins were animals (and like animals today) they led a bottom-up existence; they lived in the present and sought no major answers to any difficulties, nor made any major improvements to themselves or their environment. They were replaced by the hominins, whose top-down mental pathways had evolved to comply with heteronomous controls, which enabled them to seek answers, find solutions, and make improvements to their existence following their catastrophic loss of autonomy. They were insecure, troubled and anxious and in trying to ameliorate their situation they gained mental access to the past, present and future, constructing new modes of behaviour and social restraints that have no purpose, or appeal, for the fully autonomous animal.

Full-blown top-down neural pathways do not (cannot) evolve unless heteronomous directions are forced into the controlee's brain. There is no reason for them to evolve because the bottom-up directions, selected by evolution, are all the animal needs to orchestrate and coordinate its own safe existence. In a world without motor-control, bottom-up governance has not been found to need improvement throughout 400 million years of animal evolution.

The necessity, urgency and complexity discussed in this chapter are simply non-existent in animal behaviour. Animals do not have to defend themselves from motor-control hence they are safe, free and autonomous. Opening the mind to top-down heteronomous mental activity is an admission, by the hominins, that their autonomy has been overridden. When considered from the perspective of the controlee, what is being discussed here is a vulnerability to having to enter relationships of power, coercion, compliance and cooperation in which the once-autonomous controlee becomes a 'victim', and this is discussed more fully in the next chapter. It is only by unravelling the complexities of the strained and difficult relationship between the controlee and the controller that the necessity for top-down neural pathways can be seen with clarity – for without motor-control they have no purpose.

Endnotes: Chapter Six

Endnote 1

I have used the term 'top-down' in relation to the operation of the hominin brain, however, there are hints from the abilities of some animals and birds, when they use tools to solve problems, that their brains can function in a rudimentary top-down way. Of interest are reports that chimpanzees have been observed using a stone as a hammer to crack nuts and crows can perform complex tasks or resolve complex puzzles to obtain food.

If this is top-down mental activity it would mean that these rudimentary neural pathways were already in place and they were exploited by motor-control in the development and selection of the fully top-down hominin brain. This rudimentary top-down ability (if it exists) cannot be used by the animal to directly 'enter' the mind (brain) of another animal and to control the other animal by telling it what to do, or what not to do, because this also requires the additional factor of the ability to motor-control.

It is acknowledged that the terms 'top-down' and 'bottom-up' are crude shorthand for complex neurological activities. They are used here as a succinct and convenient means to establish some preliminary insights as to how animal and human minds differ in their mode of operation.

CHAPTER SEVEN

THE RISE AND RISE OF HETERONOMY

This chapter examines top-down heteronomous control from the perspective of the controlee. In effect, it will be the start of a troubling exploration in which we come to understand the extent to which we are forced to respond to the controls, directions and desires of others. Our individuality (personality) is determined by the ways in which we react to the myriad of conflicting demands that others make upon us, demands that invade our minds and coercively influence our everyday lives.

If we examine the contents of our minds it will be seen that a large percentage has been deliberately placed there (implanted) by others – parents, family, partners, teachers, friends, employers and the state. It is the sheer volume of these inputs that is so remarkable, for all of them influence our thoughts and actions (in gross and subtle ways) and in most cases, burden us with the unresolved tensions that result from having to comply with the demands of others. Some demands are not overtly aggressive, they may even be experienced as pleasant or gentle, but the moment they override autonomy to become unsolicited heteronomous control they violate the autonomous essence of animal existence.

We are used to the word heteronomy being used to describe being subject to the laws of a foreign government but here it is suggested that it should also be used to describe the core dynamic of our everyday personal relationships. Being subject to the control of another is a dangerous and vulnerable situation so it necessarily stirs high levels of emotion. However, for the individual it is not simply a question of the controller being in the wrong and the controlee being in the right because each of us, in different situations, can become a controller, or a controlee, thus heteronomous control always gives rise to a complicated mix of emotions in the minds of the antagonists.

7:1 The difference between the controls faced by animals and the heteronomous controls faced by the hominins:

As stated earlier there are different ways in which humans and animals are controlled; the animal simply submits to dominance whereas humans go much further and comply to the wishes and directions of the controller.

When an animal is forced to respond to a more dominant animal, the submitter acknowledges the dominance, either by retreat, by relinquishing its claims, or by submission. The submitter may not be able to eat the first fruit or the best meat, or mate with the desired female, or the female may not be able to wander away from of the herd in the breeding season. But they never fully lose their autonomy because there are always alternative actions available to them such as backing down, changing their mind or moving away, or in the case of the corralled female she may not be able to move away but she is still able to move around. Hence the 'control, influence, persuasion' of dominance is far less onerous than the overpowering (physical and mental) heteronomous control experienced by hominins.

This is also the case (see Chapter 5) when dealing with natural events such as, a flooded river crossing or a landslide blocking the path. The choice of behaviour, to back off or not, is always an autonomous decision, "I won't cross now, the water is too high," or, "I won't try to eat the carcass now, it's too dangerous," or "I cannot win this fight, I will back off." All these originally undesirable outcomes can take place without a loss of autonomy because it is in the best interests of the individual to change its mind and disregard its original preference.

However, a dominant hominin controller can impose (with the use of hands) directions that must be enacted and at the same time deny the controlee any alternative behaviour or any chance of avoidance.

In this situation the controlee's autonomy is lost or severely compromised, for example, "I am going to rape you," "You will be beaten," "You will marry this man," "You must eat this food," "You cannot leave this room," "You have been conscripted," "School is compulsory, pay attention," "Bend over," "Do as I say."

The controlee's original bottom-up programmes, which allow for flexibility in autonomous choice, are completely overridden in situations where she/he has to succumb and enact the top-down instructions of the controller.

Controllers have become so inured to the plight of the controlee that, it may be said, they override the golden rule of nature that states: "Do not progress beyond the point of submission; back off immediately this moment is reached". The ability to override submission comes in the wake of a shameful loss of empathy, thus we have become able to ignore the plight of the human controlee when she/he is caught, held and subjected to heteronomous controls.

By allowing the entry of another person into the control/governance section of the mind a Faustian bargain is struck: the controlee improves her/his chances of survival by means of compliance but in so doing compromises her/his own independence and autonomy. Compliance means that the controlee has 'allowed' the controller to gain access to her/his brain to parasitize the workings of the mind and, alarmingly, has allowed whole sections of her/his behaviour to fall under the control and governance of another.

It is the norm in the animal world for the thoughts, wishes or desires of one animal to gain little, or no, purchase in the mind of another; this is because the moment an individual becomes uneasy about any situation it can take avoiding action by the simple act of moving away. Thus, the amount of mental contact that the animal has with another is severely curtailed, it cannot escalate into 'receiving and responding to heteronomous directions' simply because it is impossible (without the coercive element of motor-control) to gain a top-down purchase into the potential controlee's mind. Therefore, animals are spared the effort and difficulty of marshalling these heteronomous inputs and therefore their minds have continued to function along (much calmer) bottom-up, autonomous principles.

7:2 Compliance – humans as agents of others:

As hominin vulnerability to heteronomous control increased it became very easy for controllers to implant their thoughts and directions into the minds of others. This human behaviour is, in effect, like a parasitic disease – once a thought, wish or direction penetrates and lodges in (infects) the mind of another then it only needs a little coercive reinforcement for it to remain there, influencing behaviour for a very long time, even perhaps the whole of the victim's life. This infection can readily be passed on to others, even down the generations, and it has the potential to last for thousands of years – for example, allegiance to royalty, belief in a god, the need to circumcise, are all powerful memes that are maintained by means of heteronomy (that is why they are not present in animal behaviour). If any of these infections were in danger of dying out other memes would appear, since they are the vehicles by which heteronomous control is maintained and carried around the world. For example, today in the Western world, you may now be able to opt out of religious rituals, but you will not (under any circumstances) be allowed to escape from the rigours of compulsory education (see Chapter 10).

To continue the disease analogy, it is hoped that heteronomous control (with all its symptoms) proves to be a disease of the 'childhood' of mankind and as the species matures we may be able to develop some immunity or better health measures to combat the damage it inflicts. At present, heteronomous motor-control cannot be eliminated because 'reinfection' takes place each time it is experienced, so we need to find a way that stops us passing on the 'virus' of motor-control, via the act of motor-control. We need to erect a *cordon sanitaire* around the act of motor-control that, in effect, limits our grasping hands and controlling commands, thus stopping the spread of this pervasive, contagious and appalling disease. However, the first task of any disease prevention is to isolate the cause, for not until the 'infective agent' (heteronomous control) is exposed to the light of examination can the process of eradication begin.

The controlee is forced, by the threat of punishment and coercion, to accept the heteronomous instructions into her/his brain and once they become implanted they need to be enacted as if they were the controlee's own wishes. Heteronomous control also has a subtler, less overt form, for once the pattern of control is established it can simply be the desires and wishes of the controller, rather than overt instructions, which prompt the controlee into activity.

Most heteronomous control does not entirely close the controlee's own space, allowing some room within which some personal needs can be accommodated. However, some heteronomous controls are extremely severe, and they restrict the controlee's space to such an extent that they are virtually indistinguishable from gross physical control. In this situation the controlee is tested to the limit and often 'breaks' under the pressure of excessive demands. This experience has long-lasting repercussions, for any gross loss of autonomy is bound to be experienced by self as being involved in a situation of the utmost danger (a situation that it should have done more to avoid). See Chapter 12 – Mental Illness.

This ability to enter the mind of another creates a very dangerous situation for it means that we are all appropriable, vulnerable to having the thoughts, wishes, directions of another, no matter what their intention, deliberately placed, inserted or planted into our own mind. When we allow this access to our mind we become infected, we become the agent of another, and for a large part of the time our body, our mind, and our physical strength is seconded, stolen, usurped, commandeered or appropriated by another.

The essence of heteronomous control is that the controlee would reject these external directions if that was possible, but she/he is overwhelmed by the power (mental as well as physical) of the controller and succumbs to compliance. It is an unavoidable consequence of control that the controlee ends up with divided and ambiguous thoughts about the situation, "I have to do it, I am doing it, but I do not want to do it."

It is inherent in the dynamic of this power struggle that the controlee's wish, 'not to do it', is universally ignored and forgotten, while the controller's wish, 'to have it done' is generally fulfilled; this is the essence of power, the controller always wins. The reluctance to acknowledge the situation from the point of view of the controlee (lack of empathy) is a wilful misrepresentation of the truth and it festers at the heart of all human behaviour causing untold mental suffering.

We not only fail to appreciate the seriousness of the situation we are in, we also fail to understand how it has come about and how far removed it is from the mental experiences of the autonomous animal. We are overwhelmed, and rendered mute, by the dreadfulness of motor-control; as a species we have never experienced any other situation. If we are to come to an awareness of our condition it needs to arise from a deep understanding of the autonomous freedom of the wild animal which has to be fully understood before it can be compared to, and contrasted with, the heteronomous restraints of the human condition. It is not possible to see or understand the tragedy inherent in the human condition (to which, to their great credit, the major religions have always alluded) before this is done. In the last resort, we are vulnerable to motor-control and we are fearful of examining this fact, so we bury our head in the sand, failing to understand the full ramifications of the tragic, unique evolutionary miss-move that has left us in thrall to heteronomous controls.

The template of compliance creates a corridor (via eyes, ears, mouth), which enables the controller to plant heteronomous controls into the controlee's mind. This means that the controlee can now be named, called, summoned, charged, criticized, imputed, declared guilty, made responsible, blamed, prohibited and punished. She/he can be ordered, directed, coerced, made to act in a certain way, subjected to pedagogy, made to repeat actions, made to believe things be they true or false, and can be made to act against her/his own will.

It is the necessity of dealing with heteronomous controls that creates the human self-conscious self, and this self-conscious self can do nothing other than comply and then reflect upon the heteronomous controls she/he suffers. The human condition is the expression of what it means to live a life continuously subject to, and under the influence of, heteronomous controls.

Every time a controlee is subject to motor-control she/

he is forced into a relationship with the controller and these relationships are mediated by emotion and language. Here lies one of the major fault lines between the animals and mankind: animals have relationships, but they are not heteronomous relationships that enforce the compliance of the controlee. Honour, pride, praise, shame, guilt, blame, dishonour and disgrace are some of the emotional techniques used by the human controllers to keep the controlee under control. One is rewarded by gifts and praise and punished by disgrace, exclusion and restriction and in the last resort subject to harm by means of physical punishment.

The sheer volume of heteronomous controls may make it difficult, or near impossible, for the individual to find time to deal with her/his own autonomous needs. The very basic autonomous priorities, such as eating, urinating, defecating and sleeping, will always create an 'urgency' at some stage, these needs do not disappear just because the controller insists upon compliance. Even those who believe it is their (heteronomous) duty to create the harshest of regimes are forced to concede this fact, for they always eventually have to provide the prisoner (victim) with food, water, toilets, and blankets unless death (rather than torture) is their objective.

All humans are prepared, waiting for the next episode of heteronomous control, even though we cannot know precisely what will be expected of us. Although we are prepared, we cannot be totally prepared because motor-control may appear unexpectedly, the controller's intentions can easily be concealed, and this makes any defence very difficult. On a hot day the girl walks home from school and the kind lady next door says, "Hello sweetie, would you like an ice cream?" Or, on a similar hot day the unmasked paedophile over the road says, "Hello sweetie, would you like an ice cream?" It is the undisclosed intent of the controller's mind coupled with the power of motor-control that makes us so fearful of each other's intentions.

Another problem for the controlee arises if controls from two or more sources are contradictory; the controlee then has the impossible task of being forced to resolve these contradictions within the unity of his/her own actions.

This dilemma (suffering) is inherent in the act of heteronomous control simply because the controller cannot be trusted to be fully cognizant of the situation from the controlee's perspective; indeed, empathy is antithetical to the essence of control. It is impossible for the controller(s) to have a full understanding because they can never be privy to the controlee's inner physiological and psychological state(s). This is true even when they come from the same background; do you know *everything* about the inner needs of your children, spouse or parents?

A self that is subjected to heteronomous governance is a fragile self simply because it does not know when a demand will be made or if it will have the strength and ability to comply. All of which has opened an evolutionary new insecurity within the hominins because the individual is no longer the sole author of her/his own acts.

Heteronomous control gives rise to a body divided against itself for it now has two (or more) masters. This means that not only self, but all the controllers of self now must be considered, referred to, consulted, reconciled, balanced and cross-checked before any decision is made to enact subsequent activity. This division of self has other major consequences, for if the control and governance of the once-unitary self can be manipulated in this way (to become an object in the eyes of the controller) then it opens the realisation that all human behaviour can be manipulated and categorized as approved or disapproved.

The first time you are forced to do something that you do not want to do there is no memory of the event that can be used for guidance. The second time you are forced to do it you will remember what happened the first time, by the third time you will have learnt what actions you need to enact to be deemed to be compliant. If you survive to reach this point, you have now become enslaved, a puppet, and with only a little more reinforcement you will remain one for perhaps the rest of your life. Your 'strings' can readily be pulled by others, who themselves were made into puppets by puppets of previous generations. We are all puppets in a world of puppets; there is not a truly autonomous human being alive, our species should have been called *Homo heteronomous*.

It is every human's fate to have a false heteronomous self, created by others, living within her/his own mind and body. Our actions and beliefs have been taken over by the thoughts and directions of others. We have been forced to comply and to take responsibility for actions that are sourced in the mind of another – the controller. These actions are not our own autonomous actions, yet we cannot survive unless we enact them, that is, we have been forced to become puppets. With age (or enlightenment) the strings may weaken for a rare and lucky few who eventually stop responding, but for most of us we live out the whole of our life not as an autonomous animal, but as a heteronomous controlee – a human puppet.

If another person can make you do things that, normally, you would not have done, then you gradually learn, realize or understand that it is also possible to make yourself (with major heteronomous prompts) do things that normally you would not have done, for example, work for self-enhancement, or lay down your life for your country. Once this area of compliance is activated it can, with daily use, be expanded until it is in full-time operation and once enlarged and alert it is ready and prepared to comply with any internal or external top-down instructions.

This neural template can expedite compliance with disturbing efficiency, as Stanley Milgram's classic experiment so clearly demonstrated (described in *Obedience to Authority: An Experimental View*, 1974). Compliance has virtually become a reflex action because non-compliance would otherwise expose the controlee to serious harm in the form of physical attack, emotional abuse, violence, restraint, rejection, ridicule or exclusion. In many situations, we now comply as fast as we withdraw our hand from the fire, or jump back when we see a snake (indeed, instant compliance, or obedience, is the stated objective of military training).

Compliance is a fearful response to authority and authority comes in many guises, it is not just the overtly aggressive, the violent or the rich that become controllers, even caring parents have authority over their children. Parents control the money, food, warmth and shelter and because they are prepared to punish when necessary and generate the allegedly 'socializing' concepts of praise and blame, they can wield considerable power over the child (I have never seen an animal praise or blame another). Parents can impose heteronomous controls, at will, for at least the first sixteen years of the child's life until she/he can leave home, and then the employer takes over as the primary controlling agent.

The desired end result of most heteronomous control is compliance. Compliance is ubiquitous and in our rush to comply we have trashed our moral compass and lost any understanding of the meaning of autonomy. When you enact the directions of a controller you become accountable to that controller, and you become responsible for the behaviour that you enact.

The controlee is forced to comply; in theory (depending on the controller) it may be possible to choose, or bargain the degree of compliance he/she will have to enact, for example, full compliance, partial compliance or minimal compliance. In other situations, there is no bargain, the controlee simply must comply with the orders. Whether you comply, willingly or reluctantly, you are still in thrall to heteronomous control: a controller somewhere, present or past, has planted the directions in your mind and is pulling the strings. Our once-sacred autonomous self must submit when faced with the overwhelming power of motor-control and that is the harsh biological fact of hominin existence. It is also a biological fact that nature had, heretofore, always programmed the dominant animal to back off at the point of intraspecific submission. Our ancestors were the first to violate this fundamental rule, so to now honour these gross acts of power and control and to praise compliance and obedience as we do today doubly violate the essential truth of autonomy.

7:3 Obedience:

Lincoln Allison's (answers.com/topic/obedience) (now unavailable) definition of compliance said, "Compliance to another's wishes may turn into a swift dependable response and it then becomes obedience. Obedience is the conformity of one person to the will of another by the implementation of that person's orders and instructions. Unquestioning obedience involves a willingness to implement instructions without exceptions. In despotisms and absolute governments, as well as in certain religious and military organizations, such obedience has been considered a virtue, but in liberal, individualistic societies it is considered morally reprehensible and dangerous."

Obedience is a word coined and defined wholly by those wielding power: monarch, state, army, church, or school. It is a weasel word where an untruth deliberately enters the language, as can be seen by a careful study of its antonyms. – defiance, disobedience, intractability, recalcitrance, contrariness, contumacy, insubordination, noncompliance, obstreperousness, rebelliousness, self-willed, unruliness, waywardness. No mention here that disobedience can be heroic, principled, autonomous, free, firm and justified.

Obedience has featured from the beginning of history; all

our problems have been laid at Eve's door for without her alleged 'disobedience' it is said, we would still be in paradise. Similarly, with the binding of Isaac (Genesis 2.2) it was the absolute sovereignty of God, over human life and death that led to the testing of Abraham's obedience, hence his willingness to sacrifice his eldest son. Note that the whole of this tragic story is dependent on having hands to control and hold, no animal behaves in this way, where is the stallion that sets out to slaughter its first-born foal?

What is so shocking here is the idea that God would find this aberrant behaviour praiseworthy but of course he (or she) doesn't, this behaviour is the perfect expression of the psychopathology of 'lost' Man – Abraham was already deeply immersed in the heteronomous control of his parents and peers for how else would this farcical idea have come about?

For the religious mind to link a top-down belief in God with obedience was a catastrophic miss-move. The essence of nature/ God is freedom and autonomy, whereas the essence of obedience is heteronomy and control. The confusion here may have come from a way of seeing autonomous bottom-up (programmed in the genes) behaviour as 'obedience', as when ducklings bond with the first moving thing they see after hatching and closely follow their mother.

There is no doubt that animals do exhibit a chemical or electrical obedience to their bottom-up programmes; one molecule of the female oestrous pheromone will summon the males of the species from miles around. In nature the individual cannot function without these chemical and electrical reactions which follow rigid rules, or laws, but this is not obedience in the heteronomous sense of the word (it is obedience to self's bottom-up programme, not to the commands of another). The bottom-up response is provided by nature so there is no reason for it to be questioned and for those who wish to think in these terms this may be described as obedience to 'God's will' for in this sense, and this sense only, one is 'obedient to' nature or God. All other understandings of being obedient to 'God's will' are meanings that have been invented by man as a means of inflicting the most extreme heteronomous control on others, a concept which is tantamount to blasphemy given that the essence of nature (God) is the freedom of the individual to express bottom-up programmes autonomously.

To be fair to the main religions, they have always maintained a core belief that addresses the issue of freedom but in failing to understand the havoc that is reaped by motor-control, they have been overwhelmed by their own need to impose heteronomous controls, so they have sinned, opting for a system of power and control rather than one that facilitates autonomy.

Back to Stanley Milgram and his experiments published in *Obedience to Authority*, 1974. He diagnosed a fatal flaw in human nature – an excessive propensity to obey others – but he did not go on to explain the heteronomous pressures that drove this propensity. Obedience in the human is much more than an extension of animal submission. Dominant animals can always get the weakest to submit but they cannot make them comply with any orders. In human society it is the ability to motor-control that gives the controller authority over the controlee. Milgram's 'obedience' isn't the fatal flaw, it is the use of hands to motor-control that is the fatal flaw because it enables the controller(s) (those in authority) to demand (induce) obedience (heteronomous compliance) from others on pain of punishment, rejection and/or exclusion. "The power to impose obedience," (*Reflections on the Just*, by Paul Ricoeur, page 92) creates its own psychopathology which is illustrated by the following quote from William Edward Forster (1818–86), an English politician noted for his Education Act: "There is no shame in taking orders from those who themselves have learned to obey."

The Abrahamic example shows how a top-down heteronomous instruction can become so deeply embedded into the mind that it must be obeyed against all other considerations. There was not a word of consultation or commiseration with Sarah, Isaac's mother. Here lies the deepest critique of the monotheistic religions, for at their heart is obedience to the top-down heteronomous moral codes of man (dressed up as God's), rather than facilitating the free flow of the autonomous, bottom-up decisions of nature. Motor-control and heteronomy have driven us from our prelapsarian state and it is the findings of science rather than the text of the gospels which will expose and illuminate this fact.

Jews and some Muslims expect their adherents to circumcise their sons when they are seven or eight days old. If it is performed without the benefits of anaesthesia (which is the usual procedure) this is an assault. The baby is forced to endure extreme pain so that the parents can comply with a heteronomous directive issued by their forefathers that remarkably has been passed down the generations for over two and a half thousand years, without once asking the permission of the child.

A better illustration of the longevity of heteronomous control would be hard to find. Had there been any real benefits to having a penis without a prepuce then it would have appeared because of natural selection many years earlier. A similar and potentially more harmful psychopathological surgery is inflicted on some Muslim girls who are forced to endure genital mutilation. The perpetrators of both male and female genital mutilation are driven by codes that pulse in every generation, which have been heteronomously planted in their brain with such intensity that they are prepared not only to override their normal parental instincts of care and protection, but they perversely deliver the child to the ceremony with pride, joy and pleasure, to be mutilated.

Do animals mutilate the genitalia of their offspring in this way? This is not God's law; it is heteronomous control pure and simple, it is control of mankind by mankind, dressed up as God's will. How could it be God's will to seek to 'improve' upon his (nature's) original design?

7:4 The role of cooperative heteronomy in human socialization:

Animal cooperation has an autonomous, telepathic, almost extrasensory element to it, as when lions set out on a hunt. This cooperation is autonomous and spontaneous, unlike heteronomous cooperation, which first requires a controller to top-down implant an idea, plan, wish or demand in the brain of the controlee before the controlee can comply.

It is relatively easy to understand extreme, coercive and heteronomous controls that are forced upon the controlee to the sole benefit of the controller(s), be they an enemy, a villain, a husband or wife, a parent, or a tyrannical state. However, there is a different type of heteronomous control that is imposed with the objective of protection rather than assault, of creating a cooperative defence where the controlee and others do gain some benefits from their compliance.

As soon as the template for compliance was in place in the hominin brain it was able to be honed to accept specific instructions from any number of controllers or any source (such as the scriptures) in lieu of an extant controller. This meant that it could also be used by the individual to form bonds and pacts and make plans to build cooperative defences with family and friends to limit, or avoid, harsher coercive threats from undesirable external sources.

These cooperative controls gave rise to our moral codes that are held in place by an agreed set of punishments for those who transgress – here lies the origin of jurisprudence. Monitoring and regulation of behaviour necessitates the close contacts necessary for control and oversight, hence we have become the most 'social' of animals. Humans have a pressing need to be social and cooperate otherwise we would tear society, and ourselves, apart with our invasive hands. However, without grasping hands there is little or no need to cooperate and little or no need for empathy.

Cooperative defences lie at the heart of all human social structures. They form the basis of the social contract and all subsequent legislation. I suggest that the reason why Rousseau was unable to find the moment when we each signed up to the social contract was because our ancestors had already tacitly included all future generations in the agreement, binding us, via our upbringing, to the rules of some form of cooperative social defence.

As future generations became socialized and educated, they became locked into the contract with or without their consent, for without some sort of agreement they would not have survived the excesses of motor-control and we, the signed-in-absentia participants of the contract, would not have been born and gone on to have progeny. So, from birth each of us is automatically enrolled and locked in to the social contract via direct heteronomous mental contact with mother, father, siblings, grandparents and neighbours who, with eye contact, games of cooing and hiding, smiling and repeated naming of the child, forge a top-down relationship without which close cooperation could not function.

Our compliance to heteronomous controls begins in these formative years, so does our blindness to the existence of these controls, for we know of nothing else. (Note the well-established mother–offspring bonds seen in animals do not necessitate this type of control; in fact, they generally avoid all overt eye contact as well as any deliberate mental penetration of each other's mind. In contrast, a common feature of human admonishment is the phrase, "Look at me when I am talking to you.")

The ability to adjust to top-down instruction has been rigorously selected, to the extent that we all now naturally develop an open acceptance of heteronomous inputs from an early age. This is selection for survival in the new and hostile environment of motor-control; we cannot avoid heteronomous inputs, so we have been selected to understand them and to react to them for, without compliance, we would be subject to the punishment of the controller and become as vulnerable as a nocturnal animal foraging in daylight.

From a year old, the child's mind is wide open, like the gaping beak of a nestling bird, ready to receive the packages of information (and controls) that are heteronomously placed into it by parents, family and friends, which steer the child into accepting the norms of social behaviour. These inputs teach the child the ways in which, historically, the family has dealt with its experiences of motor-control – conformity, obedience, aggression, reflection, education, manners, status, wealth, etc. We are open to family top-down controlling inputs because they are the first contacts we experience, and so they become the pattern and cornerstone of our ability to deal with all the subsequent heteronomous controls that inevitably will threaten autonomy throughout the whole of every human's life.

Cooperative heteronomy encompasses various, and varying, relationships, many of which become oppressive even when they begin with good intentions. It is tempting to say that heteronomous controls and instructions that are genuinely caring or educative are wholly good, but they have an inherent danger: they are always prone to escalation, and the controlee, having accepted or accommodated the first heteronomous control is particularly susceptible to having to accept further controls.

Cooperative heteronomy has now become essential for human survival; in any circumstances other than the threat of hominin motor-control this would be a totally unacceptable violation of autonomy. Here in one sentence is the desperate nature of our plight and the hollowness of our cry that we are the pinnacle of evolution, for in losing autonomy and falling prey to heteronomy we malfunction at every level.

7:5 Degrees of severity of heteronomous controls:

The degree of severity of heteronomous control varies. Some controllers may be consistently harsh whilst others may be consistently lenient. Others may be inconsistent with their controls and this inconsistency can be expressed in the treatment of a single controlee, or in the treatment of different controlees.

It is from within this unpredictably dangerous environment that humans must try to regulate their relationships and map out a relatively safe mode of existence. Some of us are fortunate enough to live in supportive stable family groups where they receive the maximum care with the minimum of controls, while others suffer appallingly, experiencing severe control and daily punishment, much of it at the hands of their allegedly close, 'loving' family.

There is also an additional complication that needs to be considered: those that are caught up in controlling situations are notoriously unreliable witnesses as to the severity or extent to which they are controlled. Many controls are designed to specifically indoctrinate the controlee into compliance and acceptance, hence, the controlee often claims to have 'chosen' to be obedient, or that her/his punishment was 'deserved'.

Thus, a trafficked sex worker may say, "I like my job, I need the money," or a child subjected to the strictest standard of obedience may say, "I love pleasing my father," or the beaten wife may say, "It was my fault, I did not clean the floor properly." Tragically, even today, many (most?) people instinctively side with the powerful controller(s) and so they do not challenge the victim's assessment, failing to see that it is made under duress, thus the controller retains her/his heteronomous power and so the lies and the injustice are perpetuated. This is not simply individual behaviour for it is mirrored in the wider arena of state governance.

Heteronomous controls can be divided into categories depending upon the severity of the associated punishment:

1. *Seriously harmful heteronomous control where punishment cannot be avoided:*
   The controlee is held past the point of avoidance and forced into submission to act out the controller's directions (forced to act as the controller wills). The controlee cannot avoid or escape from the confrontation in which the controller has unlimited power. The controlee is powerless to resist, she/he is now acting under duress in thrall to the controller. This worst grade of heteronomy involves the physical control of the controlee solely for the benefit of the controller and it is seen in all cases of rape, slavery, torture and abduction.

2. *Difficult heteronomous control where punishment can be avoided by compliance:*
   Broadly, this is cooperative heteronomy, it covers a very wide range from mild to severe control and is the most common form of heteronomy that humans experience (suffer, or tolerate). Thus, it is the level at which most human interaction takes place. It is generally thought to be helpful and beneficial (in the interests of the controlee) but on careful examination much of it is harmful. There is always a cost to the controlee for the act of compliance (non-autonomous actions) and it is usually too high. Payment for work done is a recognition of (compensation for) this cost.

3. *Mild or helpful heteronomy where no punishment is involved:* Here, the controlee readily agrees to the controller's directions because they are clearly of immediate benefit or in harmony with the controlee's wishes – "Come next to me and I will protect you," or "Come with me I will find some food." Good parenting falls into this category but it can quickly move into a more difficult category when the parents start to have plans, desires and wishes for their offspring.

As well as the severity of the control there is a second variable regarding the controlee's stance to the controller and the controls she/he is facing, and this is the degree to which the controlee is initially willing to comply with the control. When the controller applies heteronomous pressure, the controlee can:

1. comply willingly because it is beneficial or deemed to be 'correct';
2. comply because it is a conventional requirement;
3. comply begrudgingly because the benefit is dubious;
4. comply only under coercion because it is unpleasant and of no benefit to self;
5. resist if possible because it is 'wrong', terrifying or dangerous.

7:6 Duration and severity of heteronomous controls:

The experience of heteronomy can be mild, difficult, harsh or severe and the duration can range from short to long-lived.

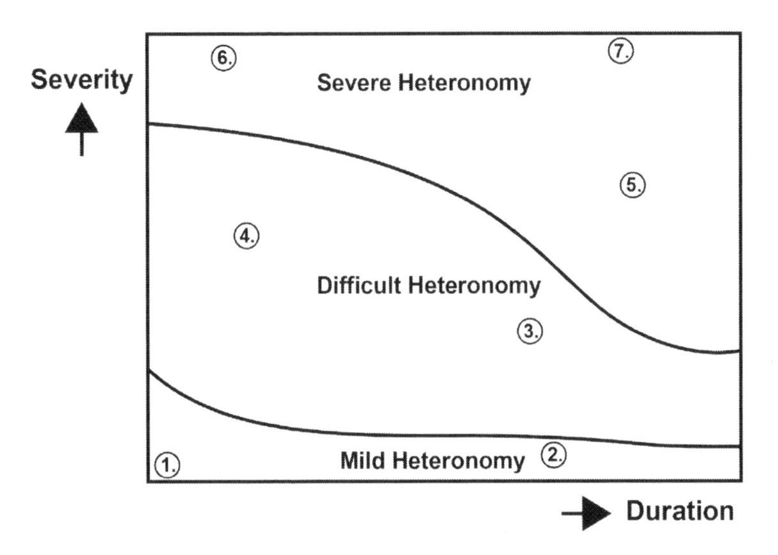

Fig 4.
Illustrating the various levels of heterononous control.

Key to Fig. 4.

1. No heteronomy (this is autonomy).
2. Mild limited heteronomy – a loving empathetic relationship.
3. Mild prolonged heteronomy – a difficult family life, marriage, life in a benevolent democracy.
4. Limited difficult heteronomy – work, school.
5. Prolonged difficult heteronomy – military service, prison, difficult marriage.
6. Limited severe heteronomy – assault, sexual assault, torture.
7. Prolonged severe heteronomy.

7:7 Length of time controller can enforce heteronomous controls:

The controller's instructions can be 'fixed' in the controlee's mind with varying degrees of permanence, depending on several factors: the length of memory of the controller or the controlee, the severity of the controller and her/his ability to inflict punishment, and the ability of the controlee to obtain assistance.

Once controls are used to create cooperative defences they have the potential to last for a much longer time than any individual control. They can be passed down through the generations in the form of moral codes and group customs and last for hundreds of years.

Short-term response to heteronomy:
In situations such as, "Do as I say, or I will hit you," the controlee is forced to keep the instructions uppermost in her mind with an immediate threat of the pain of punishment, so it is better to comply than chance injury. But as soon as the controller moves away, the immediate threat is removed, the control element diminishes, and the victim can return to his previous behaviour.

Medium-term response to heteronomy:
Medium-term heteronomy allows for a delay between instruction and punishment. For example, in the morning the controller may say, "Do as I say today, or you will regret it this evening when I return."

Long-term response to heteronomy:
The fear of brutal punishments and retribution will maintain very long-term heteronomous cooperative controls; for example, ethnic groups often fear the return of old enemies and they make sure that their children learn everything about the atrocities experienced by their ancestors. "My mother said I was never to forgive them for

the way they killed my father. Today (fifty years later), I discovered where they lived, so I went to their village and shot them."

Long-term heteronomy can also be driven by the cooperative family controls that are maintained by the fear of being weak; in this case it is vulnerability, the pain of failure and rejection that drives the controlee, rather than fear and pain of physical punishment. The social standards, expectations and needs of the family are imprinted into the mind of the child and these throb, like a radioactive isotope, driving the training, job expectations, political and religious allegiances, status and marriage of their offspring.

Falling short of these standards is a disgrace, a failure worthy of rejection. Once securely implanted/embedded in the mind there is no need for overt physical coercion to motivate the controlee because compliance is virtually guaranteed. The controlee now drives her/himself top-down to study, to pass examinations, to get a good job, marry well, buy a comfortable house, all of which provides status, financial returns and a pension, all deemed necessary to protect the individual from the disgrace, exclusion and unsolicited heteronomous controls that come with low status and poverty. Cases where the imprinted controls continue to function when the original controller is no longer present, or dead, highlight just how vulnerable humans are to having such long-lasting motivations implanted in their minds.

7:8 Some of the common ways in which humans are subjected to heteronomous controls:

Below are some common examples of heteronomous control:

1. **Toilet training** – *Who's a clever girl? Well done, who's a big boy now? Oh no, look what you have done, you are very naughty, only*

bad boys do that. What animal links praise and blame to the essential acts of urination and defecation? Does the badger say to her offspring, "You have missed the latrine, go and sit in the corner, you won't get any dinner tonight?"

2. **Table manners** – *Sit up; Take your elbows off the table; Don't speak with your mouth full; Wipe your mouth; Don't leave the table.*

3. **Social etiquette** – *Say thank you; Wipe your feet; Wash your hands; Comb your hair; Wash your face; Straighten your tie; Wear your uniform; Put that down; Don't touch that; Don't touch yourself there; Cross your legs; Don't do that; Come here.* Does an animal parent (in effect) say any of these things? They never say "Good girl" (or boy), or "Bad girl," (or boy) yet their offspring grow, prosper and breed.

4. **Compulsory education** – Note this is not autonomous learning but mandatory schooling where heteronomous controls become paramount (see Chapter 10).

5. **Work** – Animals feed themselves, find shelter and exist freely (and for free) in their species habitat. Yet, the allegedly most intelligent animal on the planet is locked into fifty years of wage slavery to provide these necessities.

6. **Religious adherence** – Attending church services, listening to sermons, saying prayers.

7. **Social norms** – Marriage, sometimes arranged marriage with a totally unsuitable spouse, sixty years of hell enforced by others, what animal gets into such intolerable situations?

8. **Military Service** – Whether a volunteer or a conscript, the soldier is forced to deliberately subject him/herself to life-threatening situations for the benefit of others (for the 'good of the country'). Without the top-down coercive power of motor-control, he/she would naturally run and hide from events like this, but obedience and duty has been so deeply implanted in his/her mind that the heteronomous controls hold sway. This example highlights the extraordinary power of coercion; the soldier can be trained to expose himself to death and injury and is given no alternative option other than court martial, disgrace or firing squad. The essential safety valve of avoidance, which is always available to every animal, is denied the human by the physical power of motor-control – he/she is held to account.

9. **Law and Order** – For some this will mean fines, imprisonment even capital punishment imposed by the judiciary for the benefit of the community. What animal needs to fine or imprison another animal in this way? In prisons additional crimes are committed often without punishment (anal and oral rape is endemic in some prisons). Here is an example of the loss of liberty ordered by one controller (the state) creating the conditions within which other controllers (prison warders, dominant prisoners) can act with impunity and so the controlee suffers repeated loss of autonomy, control upon control upon control.

10. **Vulnerability to those in positions of power** – For example, religious groups, schools, especially boarding schools, music schools, scouts, social clubs and sports clubs have, over many years, facilitated abuse by those in positions of power and authority (from politicians to disc jockeys) and the terrified victims mostly remained silent. Thankfully, the books of the

psychoanalyst Alice Miller (and others) at a theoretical level and the publicity skills of Esther Rantzen with Childline have now unleashed a torrent of awareness and increased vigilance, without which the child is defenceless.

Heteronomous control is always enforced by a controller who has greater power than the controlee. No wonder then that our minds are full of the indigestible information placed there (often without rational thought or empathy) by others; children's minds are particularly vulnerable, and they are regularly used as garbage bins for parental rubbish. When the controller decides to enforce his 'correct' way of behaviour the controlee is denied avoidance or any alternative course of action.

Often the only way for a controlee (especially a child) to modify this imbalance is by so-called 'bad' behaviour, which is behaviour that temporarily denies the parents their desire for control. This in turn allows the parents to ratchet up the harshness of the control; backing down is an unlikely option for a controller who is being challenged by a subordinate.

7:9 Assessing heteronomy:

Motor-control and heteronomy are hidden in plain sight, which has made it easier for us to come to accept heteronomous control as a perfectly normal part of human existence. As a result, when heteronomy is under discussion, there is a great difficulty explaining, judging and criticising its harmful effects. It is also a 'hot-spot' of emotion, which makes it difficult to focus on the underlying principles.

So first, we need to recognize the extent to which we are in thrall to heteronomous controls and then acknowledge that all other motile organisms do not lose autonomy except at death. This fundamental rule was not breached until the hominin hands were freed from the restraints of the trees. We are now so vulnerable to coercion that the hominin female cannot establish anoestrus. It is not just that we do not acknowledge these facts, we repress any knowledge of their pervasive existence and instead indulge in myriad displacement activities. We praise human intelligence, we are obsessed with non-breeding sexual activity, we specialize in training others to comply and obey, we prepare for and participate in warfare and we spend much of the day in the pursuit of wealth, all to keep our minds from facing the underlying problem of our lost autonomy.

Second, it needs to be acknowledged that the hominins (particularly the females) are extremely vulnerable to motor-control and only by participating in cooperative defences with others is it possible to survive the worst excesses of coercion. However, these cooperative defences depend upon their own heteronomous controls. Thus, we have a situation where we trade acceptance of the controls of cooperation, for protection from the more threatening controls of the overt controllers. This is the tacit agreement of the social contract. These cooperative controls and responsibilities have coalesced into an adherence to a never-ending set of social norms and requirements that are far removed from an animal's spontaneous (autonomous) expression of its bottom-up programmes.

Third, we are all victims of heteronomous motor-control, but it should also be remembered that each of us inflicts heteronomous controls upon others. For every heteronomous act that the controlee must enact there is always a controller somewhere, at some stage, instigating and maintaining the controls.

Fourth, once all facets of motor-control are seen and understood as a serious transgression of nature's norms, then it should be possible for society to gradually reduce the burden of heteronomy that each of us is forced to carry. The utopian objective would be to eliminate heteronomous controls altogether, letting the human mind return to its animal, bottom-up base, organizing purely autonomous governance.

7:10 Assessing the percentage of time spent under heteronomous (rather than autonomous) control:

A behavioural decision made under conditions of heteronomous control is not a truly free decision, but a choice made in bad circumstances – always a second-best choice. This is a 'subjugated freedom' where choice is made from a position where the individual is totally unaware of the experience of, or the meaning of, making a truly autonomous decision, living always in an environment pervaded by heteronomous controls.

When an action is performed under heteronomous control it is the controlee that must physically enact the directions of the controller and so an action that has a totally heteronomous source is incorporated into the outputs of the controlee. Confusingly, the action is then attributed to the controlee, even though his/her actions are directed, either overtly or covertly, by the wishes of another. This misattribution is also accepted by the controlee, who usually believes that because she/he has enacted the behaviour then it is her/his own behaviour, without acknowledging, appreciating, pointing out or insisting that the primary source is the controller and the controlee is acting under heteronomous control. ("I didn't want to do it, he told me to do it and now he says that I am not doing it right.")

If for every action the original heteronomous input could be calculated and flagged up, say, 50:50 or 20:80, then the non-autonomous content of human behaviour would be clearly visible, and it would be seen that animals act autonomously virtually all the time and humans act heteronomously virtually all the time.

Children are under direct heteronomous control most of the

time. “Come here,” “Go to school,” “Wash your face,” “Don’t do that,” “Eat your tea.” The command “Eat your tea” is interesting because the child needs to eat, so when Mummy says, “Come and eat your tea now,” some section of the child’s mind is under strict top-down heteronomous control but another section under bottom-up control needs to eat and, in this situation, the act of eating will result from a mix of heteronomous and autonomous responses. Similarly, for the worker when the boss insists on the working of overtime, this is a strict, perhaps arduous, control but the worker will also welcome the extra money to pay some bills.

The controlee may experience simultaneous heteronomous controls from several controllers, who may issue conflicting instructions. This can cause considerable stress in the controlee where his one course of action cannot please all the controllers. A child will have to absorb inputs from mother, father, grandparents, friends, neighbours, schoolteachers, priests, sports coaches. So, the child’s behaviour is the result of a mix of inputs from various controllers, none of whom necessarily have the same information available regarding the child’s physiological or psychological state, nor anything approaching the child’s internal information when they make these decisions. Yet if the child is to develop the skills and defences to live in a world where he will be subject to motor-control, this mass of conflicting information must be dealt with in some way or other, and much of it will have to be memorized for future reference – hence the need for an enlarged brain.

As well as spending a different amount of time under the thrall of heteronomy, controls may involve a different number of controllers and controlees, for example:

1. One controller directs one controlee.
2. Two to ten (say) controllers direct one controlee.

3. One potential controlee organizes his/her own defence with, say, up to ten other potential controlees assisting in that defence.
4. One king, president or leader heteronomously controls several million subjects.
5. Ten to thirty members of government control an electorate of millions.
6. One child has heteronomous inputs from up to (say) thirty others – parents, grandparents, friends, family, and schoolteachers.
7. One adult has heteronomous inputs from up to (say) one or two hundred others – friends, family, work colleagues, public figures, or from fictional characters in television programmes, books and films. So that during a lifetime the average individual may be influenced by the wishes, desires and directions of many thousands of others, many of whom she/he never met.

7:11 A scale of coercion and heteronomous control:

The levels of intraspecific heteronomous coercion experienced by different groups can be compared. Here they are placed on a scale of diminishing freedom, starting with wild animals that always maintain autonomy and ending with the victims of rape, murder and capital punishment.

1. All wild animals are autonomous and free from heteronomous controls.
2. The enlightened human (those rare souls) who may have managed to release themselves totally from the burden of heteronomous controls.
3. The bushman, aborigine, the Amazon forest dweller, the

wanderer, the loner, the layabout is possibly subject to the least heteronomous controls.

4. The average wage-worker, fifty years of direction, supervision, time-keeping and toil.
5. The religious adherent whose actions are directed, often to the letter, by codes of behaviour written perhaps thousands of years ago.
6. The highly trained athlete. Driving his/herself to perform in an unnatural way, the training schedule allowing no respite until the competition is over. (Predators, the athletes of the animal world, do not test themselves to this excessive extent, even though their next meal depends upon the chase.)
7. The volunteer soldier. Eager, disciplined and prepared to sacrifice his/her life for their country. The soldier's drill is designed to achieve a strict adherence to the instructions of a senior officer – this is compliance honed to perfection.
8. Orchestral performer. Of all the voluntary activities this is probably the most controlled, for two or more hours the musician's movements are rigidly dictated by the score, the commencement and duration of every note must be precisely timed and executed. There is little room for any other movement, not a cough, or a scratch and certainly no departure from the chair.
9. The child in compulsory education, fourteen years in which attendance is enforced by law.
10. The conscripted soldier.
11. Abduction and enslavement. Including adults and children taken to work as slaves or as sex slaves.
12. Rape and torture victims. Whether they stay alive hangs in the balance, if not they become murder victims.
13. Finally, those condemned to death (by a legitimate or illegitimate authority) play out the most extreme

> heteronomous control where any defensive fightback is denied. The adulteress, about to be stoned, and the murderer, about to be hanged, are swept up in an entirely heteronomous chain of events, commands and security that overrides any attempt to escape. The murderer walks to the scaffold unaided. The adulteress, in the last frail autonomous move, raises her head above the level of the earth being shovelled into the pit into which she has been thrown. In so doing she endeavours to stay alive for another few minutes, only for stones to rain down on her head until she succumbs. Her final reflex is to urinate and defecate; the body's autonomous, bottom-up, defensive system coordinating the final seconds with the best and only adjustments it can make to aid survival. Execution is the final act of control; it is designed to control the controlee by means of elimination. (1)

Once heteronomous controls are planted in the mind of another they can be extraordinarily tenacious, sometimes impossible to dislodge (that is the controller's desired aim). To illustrate I shall use a true (but disguised) situation.

A recently qualified medical doctor, from another country, met a female doctor who gradually became more than a colleague. They lived together, became engaged and started to arrange the wedding. Far away his parents became distraught as they had always wanted their son to choose a wife from the country of his birth. After several months of turmoil, the son broke off the engagement. After a few more months of turmoil he tragically committed suicide.

First it is important to acknowledge the pain and suffering experienced by all parties in this tragic series of events in which there were widely differing views as to what was, and was not, the best course of action. But, for this discussion, the area I wish to concentrate on is that the son listened to and was heteronomously influenced (controlled) by his parents (by the apparently relatively minor control of displeasure).

In this tragedy it can be clearly seen how the wishes, desires, needs, pride and pleasure of the parents are bound up in the life and actions of another person – their son – and how his view of right and wrong is bound up in his parent's wishes, beliefs and desires and how he has been primed (imprinted) to put those wishes above his own.

This was a mature doctor, highly educated with knowledge (no doubt) of history, philosophy, ethics, psychology in addition to his medical training, which for all doctors is designed to enable them to deal with extremely difficult situations and conflicts of interest. Yet, despite having this extensive body of knowledge, and these skills at his disposal, he was unable to override the heteronomous controls of the parents (that had been placed in his mind and regularly reinforced from an early age) relating to the most individual, personal and important decision of his (and another's) life – the choice of spouse.

This illustrates how tenacious heteronomous controls can be once they have been planted in the mind, for on the surface it would seem to be reasonably easy for him to say to his parents, "I am sorry, Dad, I know that this will be a great disappointment, I understand that you hoped (and I thought) that I would marry a wife from my own country but I have fallen in love here and we will be getting married next year." Equally, one would have thought that the parents, on seeing their son's happiness, and resolve, would have given their much-loved son their blessing, but tragically this was not to be.

The doctor, mature, qualified, and of independent means, made a choice and was supported in that choice by his wife-to-be, her family members as well as his and her friends and colleagues, yet he then overrode his choice, rejected the woman he loved and who loved him, for fear of upsetting his parents. There was no physical control or injury, no beatings, no abduction, no confinement; it was a straightforward case of transgression of an expected compliance to a heteronomous control. With great sympathy, I rest my case – heteronomous controls can be extraordinarily powerful, and we grapple with them at our peril – no animal comes anywhere near to engaging in such complex relationships.

We are, to a greater or lesser extent, all in thrall to heteronomous control. It is difficult to determine the percentage of one's behaviour that results from the wishes, desires and instructions of others but for a social person it will be considerable. Indeed, without these heteronomous ties we are thought to be unsociable, difficult and perhaps in some way needing therapy. However, without heteronomous control we would be autonomously self-sufficient, our minds would lose their unease and uncertainty and we would experience the equanimity and harmony that animals take for granted.

A wider view of how we deal with heteronomous control is considered in the next chapter.

Endnotes: Chapter Seven

Endnote 1

A critic may be concerned by the inclusion of 6, 7 and 8 (in the list in section 7:11) on the basis that the individual freely chooses to compete, enlist or perform. However, I can see no valid reason why she/he should autonomously choose to make such an excessive or risky commitment, (often physically and mentally overtaxing); no animal does anything like it, no animal sets out to perfect their abilities or develop new skills. No animal is rewarded for its performance; do they say, "The impala with the longest leap wins the gold medal"? However, the drive for excellence, status, money, as well as the dreams of parents and others, are likely to play a large part in these ambitions and these are all classic expressions of a controlee caught up in a web of heteronomous control.

In the case of orchestral performers, they not only commit themselves to very many hours practice (starting when young) but the objective of this practice is to strictly 'lock' themselves into intricate dictated patterns of timed movement (perhaps for two or more hours). The artist, or soloist, does contribute an element of individuality to the performance but that can only be expressed within the very rigid confines of the score and even this freedom is not permitted (or tolerated) for the supporting musicians.

The athlete similarly trains for very many hours but the outcome, although constrained, is more flexible than the musician, to run say, 200 metres in his/her own style within a single lane on the running track. Nevertheless, excessive training and organized competition are not bottom-up activities (like hunting or migrating) hence the inclusion of athletes in the above list.

CHAPTER EIGHT

STRATEGIES TO DEAL WITH MOTOR-CONTROL AND HETERONOMY; ACTIVE AND PASSIVE DEFENCES

Having laid the foundations the hypothesis that states that the use of hands gave rise to motor-control which in turn led to coercive mating and heteronomous controls, I will now examine some of the additional difficulties that have arisen as a result this major disruption in hominin relationships.

As the hominins became victims to heteronomous controls their brains were tasked with balancing an essentially unresolvable conflict. The controller and the controlee's opposing interests must find some working interaction within the single entity of the controlee's brain so that she/he can regulate and enact behaviour that was sourced in the controller's mind.

This tension has two facets, one passive the other active. The passive response facilitates a greater acceptance of heteronomy (tolerance of motor-control), this is an integral component of hominin domestication and is fundamental to the adoption of compliant and cooperative defences.

The active defence is where the individual discharges the energy generated by the fear of motor-control. This tension seeks discharge either by enacting the heteronomous instructions once they have been implanted in the controlee's mind, and/or by attempting to create defences to avoid being subjected to these controls.

Hence, the enlarged brain must find ways to discharge the high tension of anxiety that arises from being subjected to motor-control while at the same time it must restrict, dampen, or repress activity to accommodate the passive domestication that is required to live alongside the controllers. In the modern world, work, sport and travel have become safety-valve activities that achieve this difficult balance, discharging the tensions of control within the parameters of domesticity.

Wild animals living their everyday lives in their own environmental niche are placid, equitable and content. That is, when at rest, their minds have a tranquil mental base from which the active programmes of, say, eating, fighting and flight are set in motion. When these programmes have been discharged the animal's mind returns to its restful base to await the next priority action. When predators kill their prey, they are not angry but simply acting efficiently; even when injured or sick an animal does not become agitated but quietly awaits the outcome. It is only when caught, or held, that the animal 'goes wild' that is, it becomes frantic for release. Feral animals are not just intolerant of motor-control, they set in motion defences well before any unsolicited contact is made by becoming extremely wary of even having the space around them (their flight distance) encroached upon. In the unlikely event of capture they react violently, seeking any way to escape and often die from shock unless they are tranquillized.

So, I suggest, that when the early hominins, with their well-established animal fear and flight mechanisms were first motor-controlled, it created a serious problem. Violent reactions would have been common, but they would not have been conducive to survival, so reactions became more restrained. The hominin controllers were not predators in the sense that they sought the death of the controlees so those that did survive were forced to coexist with the controllers in social groups. This required a selection for passivity and domestication, where the hominin became submissive and tolerant of control. But there is a difficulty with this defence because passive acceptance exposes the controlee to more and more demanding, damaging and lengthy controls.

Here is the paradox: hominins have inherited a placidity that enables them to comply with the heteronomous controls, yet they also must release the high anxiety levels that result from being vulnerable to those controls. This underlying agitation is discharged in displacement activities that constantly seek, explore and test any avenue that may lead to an improved defence (hence the division in this chapter between passive and active defences).

8:1 Passive or submissive defences, tolerance of motor-control:

Domesticated animals are more tolerant of motor-control than their wild cousins, since they have been artificially and rigorously selected from more placid strains. Hominin evolution has also rigorously selected an ability to withstand motor-control, and this has resulted in a 'self-domestication', where only those capable of withstanding control and enacting heteronomous directions would have survived.

Judging by the violent reactions of wild animals when they are captured, many female protohominins would have perished, being unable to survive the trauma of being held and coercively mated; those that did survive would succumb to emotional and behavioural difficulties that in turn would make them less fit for breeding, leaving the few placid (less perturbed) individuals to become the progenitors of the new hominin species of primate.

To enter the mind of another the controller must threaten, control, damage or punish the controlee otherwise his desire for control and governance would easily be ignored (eventually, when top-down controls and domestication were fully established, benefits and inducements such as food, housing and protection would have been used in lieu of some of the more physical threats).

Initially, motor-control is resisted by the controlee because it is dangerous and hurtful but as domestication takes hold an unavoidable acceptance is inherited so that, like domesticated animals, we seem oblivious to our plight. Under the threat of coercion, a bargain is struck in the controlee's mind to the effect that, "I will do what the all-powerful controller wants in the hope that it will reduce the extent of the harsh physical and psychological damage that I may suffer now, or in the future." Being unable to resist has meant that the hominins have lost their independence and they have become passive, weak, submissive, amenable, cooperative, compliant, acquiescent, consenting, tractable, malleable, tame, docile, meek, manageable, willing and servile compared to the wild animal. Wild animals resist controls, they do not cooperate, they do not allow another animal any control over their body or allow direct access into their mind, they do not consent to heteronomous governance, and they always take great pains to maintain their independence (autonomy).

This hominin tolerance of motor-control went hand in hand with the anatomical, physiological and behavioural adaptations associated with domestication, (such as; the reduction of canine teeth, an increase in the range of facial expressions, the growth of breasts on non-lactating females, loss of overt oestrus, defensive home bases, food sharing and pair-bonding) set out in Chapters 2 to 4 of this book. These changes were dependent upon having a brain that could accept and regulate the trust and cooperation (passivity) that are encompassed in the term 'domestication' while at the same time be able to marshal and enact the heteronomous responses demanded by the controller.

Cooperation, bonding and social cohesion are markedly improved if family, friends and neighbours can empathise with each other. It is the ability for empathy, compassion and cooperation that enables groups to develop such things as food provisioning, communal shelter construction, health care and assistance, fostering, communal gatherings, communal judgements, rituals, burials, physical defences, and to act altruistically.

These are the ways in which cooperative defences have created a reliance on group support and cohesion, which helps protect its members (particularly the females) from external controls. Animals, who live in a world where their autonomy is rarely threatened by members of their own species, do not have a need to show empathy, consideration or compassion (except at the point of submission or within the maternal bond) so these traits were not extant when the early hominins started on their downward spiral to becoming the only species that has found itself subject to motor-control.

8:2 Passivity associated with child carrying – hairlessness and adolescent delays in maturity:

An early challenge for the hominin top-down neural circuits arose out of the difficulties experienced by the children of the first bipedal mothers, difficulties that were compounded by her subsequent loss of body hair. These problems necessitated a new tolerance to motor-control to ensure that the essential maternal bonds, and familial interactions, were able to be established and maintained.

Here, we need briefly to return to the apes. As we have seen,

(Chapter 3) young apes happily cling onto their mother's body hair and the mother happily allows the offspring to hitch a ride, leaving her hands free to move safely through the treetops. At rest the young can move away from their mother under their own free will, similarly they can return to her and cling on, seeking protection at times of stress or danger, the mother acting as a constant refuge. Primate mothers may cradle their young, but they do not hold on to them, it is the young that hold onto their mothers. Sometimes the mother will snatch their young away from danger, but they always quickly let go when the offspring is safe, allowing it to adopt its own 'cling on' position on the mother's side or belly.

The hominin loss of body hair, together with bipedalism, has considerably complicated this functioning primate system of maternal care, which is vigilant in its desire not to compromise the autonomy of the child. Basically, the hominin baby cannot hold on to a hairless mother as it has nothing to cling to, but like any other young primate it needs to seek refuge on its mother, so now the mother must do all the holding. This unique situation is fraught with difficulty. The child must be able to suspend autonomy (that is, reduce its wishes, desires and curiosity) so that the mother can pick it up and hold it and put it down as and when she (not the child) chooses. At the same time, the mother must try to be sensitive or empathetic to the wishes and desires of the child, even initiating wishes and desires in the child to assist it by encouraging essential developments such as learning to walk, encouraging play, or exploration.

Being able to be picked up and moved around in this way has resulted in several behavioural and developmental changes that have delayed independence.

First, the altricial period (before motor function and autonomy is fully established) has been extended to produce a period of about two years in which the baby lacks full mobility. Second, the period in which autonomy can be suspended (as when kittens are carried by their mothers up to the point of weaning) has been extended, so that the hominin child can be carried until they are too heavy for a parent to manage. (Wild chimpanzee mothers are respectful of their offspring's autonomy, they are careful not to pick up and carry their young but wait for the young to 'chose' to hitch a ride but pet and zoo chimps can be carried if reared as captives and will solicit to be carried. Professor Tetsuro Matsuzawa of the Primate Research Institute, Kyoto University, says, "A chimpanzee mother never scolds her baby. She never hits her baby. She does not force her baby to behave according to her will.") Third, the ability to solicit motor-control has been developed so that children (and injured, pseudo-injured or rich adults) can seek to be carried by others. Fourth, a period of adolescence occurs allowing the period of dependency to be extended still further so that full independence is deferred until the individual is perhaps sixteen to eighteen (or perhaps even twenty-five) years of age when she/he is strong and experienced enough to partially defend autonomy, negotiate the heteronomous demands of the controller and be sufficiently domesticated to negotiate the social mores of their group.

The reason that it is inevitable that the passive (submissive) defence is adopted by all hominins is because the child always experiences the first acts of motor-control from within its own family. These first constraints on autonomy are introduced at a young age and in a gentle manner by the parents. However, they are experienced in the face of the disproportionate strength of the parents, and these are repeated on many occasions every day of the child's life. Whether it is prolonged agonistic buffering or corralling of the children for safety, or lifting and carrying them when on the move, or the insistence of social restraints, they all inculcate a submissive acceptance of control in the mind of the child.

Until recently these pervasive controls continued for most

females (and the weaker males) throughout the whole of life, only the strongest or richest males gaining something near to enjoying freedom of movement. But note, even those who gain their 'freedom' are still burdened by the heteronomous elements in their mind that have been deliberately planted there by parents, teachers and officials.

Throughout history it is this ingrained submission that has made it possible for emperors, kings, states and religions to gain the compliance of the people to their imposed laws, rules and exhortations. In a world subject to the threat of motor-control some form of social contract is essential otherwise the permanently receptive female has no protection and no safe area in which to raise her children. It is the period of prolonged dependency that allows others (grandparents, aunts, sisters) to assist the family unit but it is the need for cooperative defences to heteronomy that drives this intervention, for if the hominins behaved as animals do, then we would autonomously go about our own business, always taking care of our own (but not another's) needs.

While discussing this area of care and protection, we can see that there are alleged safety benefits to be gained from the control of our children. Aids to heteronomous controls, such as cots, playpens, highchairs, car seats, playschools, as well as verbal and physical restraints, all undoubtedly protect the child from many of the man-made dangers that are placed in the path of a child as it grows up in the modern overpopulated world, but they do so at the expense of the child's autonomy. If it were possible to draw up a balance sheet that also registered the psychological price that is paid every time autonomy is lost, then our assessment of the benefits would alter considerably.

8:3 Inhibition of the male sexual drive by domestication:

Possessing hands that are capable of grasping mean that it is anatomically possible for males to rape at will, but domestication means that a large majority have learnt not to do so even though the female is permanently receptive.

This inhibition of the male sexual drive (mentioned in Chapter 3) (which includes a reduction of testosterone levels compared to apes) has reduced the excesses of coercive mating and has been crucial for hominin survival. In addition, the adoption of pair-bonding and the consequent reduction in promiscuity has also placed restrictions on the males. The males would slowly have become more compliant to laws enforced by family and society because they feared punishment or exclusion that, if enforced, would restrict their ability to breed.

These restrictions, now enforced by means of the law, are interesting in that they result from society's collective, cooperative use of motor-control against noncompliant individuals. Also, many males will have come to empathize with the pain and distress of the coerced female (especially their mothers or other close female relatives) all of which will have had (has) a restraining, restricting, ('castrating') effect on the original, instant, hundred per cent mating response inherited by the original hominin males. (That is, except in times of war, when law and order break down and the males often revert, abandoning their domesticated self-control). In a stable, robust society, male self-restraint is heteronomously encouraged, and this provides the female with some space in which to establish and maintain a period of quasi-anoestrus within her underlying permanent receptivity.

Thus, a remarkable inversion has occurred; motor-control is now used to *improve* safety. This can be seen in the strict social rules associated with marriage (backed by the threat of punishment or exclusion) such as, "Keep away from my daughter," "There must be a ceremony first," "I name you man and wife," these rules have emerged to provide protection for the permanently receptive female, although today, where the female benefits from rigorous legal protection, some of these rules are now seen as restrictions.

It is within these cooperative social controls that morality is nurtured; animals simply bypass morality because they never need to protect themselves from heteronomous coercion, and so they do not find themselves in a position where it is a benefit to tell each other (top-down) what to do, or what not to do.

8:4 The active cooperative defence to motor-control:

The submissive defence has helped resolve one set of problems but there is another difficulty which the hominins need to regulate and that is the discharge of energy caused by the anxiety levels produced by the constant threat of motor-control. Tension seeks discharge, and this gives rise to a continuous, active (occasionally manic) search for, and maintenance of, a so-called 'satisfactory' defence. This restlessness within human history can be seen when contrasting the crocodile, whose behaviour has not changed for millions of years, to the human, whose lifestyles are constantly changing from one generation to the next.

The tension that arises from the threat of motor-control and from the female's permanent receptivity seeks discharge in various forms, such as conflict, work, travel, sport and sexual and religious activities. This energy release can be seen throughout the whole of hominin evolution, starting with the search for safe areas in which to live. Then, keeping ahead of rivals by constantly moving on, in the 'trek out of Africa', and later, as population density increased, the building of defensive sites and cities. As the arms race between the controllers and the controlled intensified there was a constant need to construct and use new weapons, and to create buildings, artefacts, ornaments, clothes, fashion, entertainments to exhibit wealth and status.

Ultimately, no peaceful conclusion can be achieved by discharging the energy that comes from the tension of motor-control for only the absence of motor-control can do that. Animals simply do not have this reservoir of readily available energy to enact these ill-conceived projects. We may think that this is due to the mental limitations of animals (or the superiority of humans) but a more accurate assessment is that animals do not have the inherent human vulnerability to motor-control, and so they do not have the resulting levels of tension to discharge.

There is also an added complication: it is understandable, even beneficial, if an individual has the energy available, to plan and enact elaborate defences for her/his own safety, but to be so vulnerable that this energy resource is available in a neat 'body package' that can readily be heteronomously controlled and used by others is a major species weakness. Predatory carnivores access a cache of energy when they eat the meat of a prey animal, but they can never make that animal work for the benefit of the predator and use the energy of that animal while it is still alive. Energy that can be commandeered (harvested) in this way and used for the benefit of unrelated others is an extraordinary resource that can do no other than tempt exploitation. No animal (except those animals domesticated by mankind) can be forced, tricked or persuaded to work for the benefit of others. This is not a cause for a celebration of human exceptionalism; it is an area of profound disquiet. Animals do not toil for the benefit of unrelated others because none succumb to heteronomy, there simply is no way to get into their mind to make them perform such activities. (Certain insects such as Slave-maker ants do take over colonies and enslave the workers, but this is done with chemicals, pheromones or food intake, not by physical control or heteronomous mental contacts.)

Human controllers find it relatively easy (by means of overt, or covert, threats) to arouse and use the energy of the controlee, for example, "Load that cart," "Build that wall." Having an energy reserve that is so easily available (vulnerable) as this, is an extremely odd feature, indeed, it is a unique situation for a mammal. Even more remarkable is that these things are taken for granted especially as the controller always gains hugely by hijacking the energy of others and using it to his own advantage. Wild animals do not lose autonomy and so compliance with the controller's wishes is outside of their experience, yet *we* look down upon *them* (what they think of us is not recorded).

Fear of motor-control generates a great deal of energy and anxiety and so does the knowledge (processed in different ways by the male and the female) that the female is permanently receptive. Knowing that all females are permanently receptive sets up a major tension in the male who is in a constant state of uncertainty, for is the female available or not, is she on or off heat? All of which requires an enlarged brain to understand the restraints and complexities compared to the simplicity of the oestrous pheromones.

For the female there is also a considerable tension that needs to be dissipated, for permanent receptivity means that every day she must be prepared for unwelcome (or coercive) mating. Having lost the ability to establish anoestrus she no longer experiences the long safe periods of time where she can relax in the knowledge that there is absolutely no possibility of mating.

The tension of this constant preparedness gives rise to a restlessness that (as with the male) seeks discharge in a phenomenally wide range of displacement activities that have become traditional human behaviour. Mankind makes a firework display of its existence, pyramids, towers, castles, skyscrapers, aeroplanes, arrows, bullets, shells, and rockets all reach into the sky in a grandiose display powered by the energy that is produced by the fear of motor-control. Similarly, work, farming, manufacturing, exhibitions of fashion, status, or wealth, military and royal ceremonies, training and education all require an excess of energy for them to be accomplished or enacted.

In addition, there are gatherings, spectacles, parties, festivals, entertainments, singing, dancing, sports, fairs and markets; all exploit the tension that comes from the threat of coercion. Does any animal wear a fancy-dress costume and manufacture alcohol, so they can lose consciousness? Do animals barter goods, sell ice cream or beef burgers? Do animals flock to see their pope, queen, president, film stars, or singers and request their signatures?

For the hominin species, this is a colossal release of energy that is far more than the relatively simple needs of basic survival (eating, sleeping, mating, and temperature regulation); only a major dislocation of the species' safety/danger scales could result in such a remarkable phenomenon.

8:5 The thinking defence:

Thinking is a specific form of the active defence. The need to discharge the energy that arises from the anxiety that results from the threat of motor-control is at odds with the passivity required for domestication; thinking is very active in the sense that it causes the enlarged brain to consume large amounts of energy and it fulfils the need to plan defences to the never-ending threat of motor-control, but it is non-active in the sense that it can be undertaken while sitting still.

In 1958, Lawrence Freedman and Anne Roe wrote in *Evolution and Human Behaviour*, Yale University Press:

> "Frustration may well be one essential prerequisite for the sharpening of consciousness, the awareness of self, and the development of thinking. Absence or delay in fulfilment of

> a need may result in tension, which stimulates heightened consciousness and the beginning of a self-image. If the available repertoire of motor action fails, then certain inner processes may successively or simultaneously occur."

Freedman and Roe probably did not fully understand the complexities of heteronomous motor-control, but they seem to be the first scientists to suggest that the organization of the hominid brain could be affected by problems associated with intraspecific restrictions of motor activity.

Motor-control halts the original intended autonomous activity, and so it creates a space where the next move is uncertain. This delay in the response activity produces an intervening thought process, which essentially is a checking process, a seeking process, an understanding process, and it marks humans out from other animals whose thinking is in harmony with their spontaneity.

With humans, the heteronomous thoughts, wishes, desires and intentions of the controller must be considered before the controlee makes a response and the controlee's autonomous responses now must be repressed in case they, one way or another, provoke or upset the controller(s). As external controls increase, internal controls to organize and marshal these demands rise commensurately, demanding extraordinary gymnastics within the mind of the controlee. At present, we are proud of these human abilities but an understanding of their troubling, underlying origin will eventually give rise to a reassessment.

Defence to motor-control cannot be wrapped up into neat instinctual programmes because the controller's wishes are unpredictable, arbitrary, and above all, heteronomously driven. Encephalization has proved to be the only way to deal with this open-ended problem. This is not to say that animals do not think (have conscious thoughts), it is simply that they do not have such intractable problems that they continuously need to think about, search and consider their predicament in the way that humans do because they are not subjected to the almost intolerable levels of coercion and restraint that humans face daily.

Finding ways around these restraints (or inflicting them on others) takes a prodigious amount of brainpower to plan and implement. This growth in neural circuitry, coupled with the bipedal restriction on the size of the pelvis, has led to the human cephalo pelvic disproportion that is likely to create a problem every time a mother gives birth. For this to remain the case after several million years of evolution points to the fact that both the neural circuits and bipedalism are essential elements in hominin survival and the reason that this is so is because both are indispensable in the never-ending battle to contain the threat of motor-control. At base, human thinking is a quest for safety and the dynamics of this quest means that it will continue until safety is achieved.

8:6 Unceasing mental activity:

As you sit quietly in a chair all manner of thoughts enter your head. Shopping to get; the plight of Aunt Christine; the garden is too big; Kelly's wedding; I'm too fat; petrol is so expensive; those terrible people over the road; what would Father have thought of this government? I must stop drinking so much; soon be on holiday; what shall I do about my pension? All these thoughts flick in and out of consciousness in a seemingly haphazard way. These are the wandering thoughts that the Zen master stills as he quietens his mind.

Our minds have a constant radar 'ping, ping, ping' – this thought, next thought, yet another thought, continuous thoughts constantly flashing into and out of the mind. We inherit this radar, so it must have been of use to our ancestors in their need to mitigate the worst effects of motor-control. Those without this radar would have just sat there like the Dodo unaware of the danger that was about to strike. The incessant alertness of this radar tests to see what may come up, good or bad; it is testing to see if there is anything that catches the attention, anything amiss, anything that may be of use in the incessant search for improved defences to motor-control. How will this affect me and my relationships? What price may I have to pay for any changes? How shall I act in the future?

This is the inevitable consequence of top-down mental organization that has evolved to deal with heteronomy; one can never be sure, nothing is resolved, and nothing is finalized, hence the constant top-down activity. The human mind just keeps throwing up anything for consideration, that's the principle – throw up anything and see where it leads. Alertness honed to perfection but alertness for a danger, not experienced by the animal, and not fully understood by the human. We are born with this constant unease; we cannot halt it because it is the inevitable result of being vulnerable to the ever-present threat of motor-control. Animals do not fill their minds in this way; in this respect, their minds are empty, they do not top-down scan their mind hoping to find answers to their predicament because there is no predicament – they live autonomously.

This basic radar of anxious monitoring is so essential that it has been incorporated into our inherited programmes; we do not top-down learn how to do it, we cannot switch it off, it appears in our mind bottom-up, like breathing, urinating and defecating; it is just there. The programme to use top-down thoughts in this way is just there because it has been selected, it is needed, it is driven by evolution, it is safer than the lack of searching and monitoring that went before. Faced with motor-control there is no choice but to seek out difference; the human does not have the luxury of the animal mind that functions perfectly without any top-down enhancement of its bottom-up programmes.

This use of top-down neural pathways has resulted in the expansion of the hominin's ability to lay down and recall detailed memories. It has enabled the development of speech, words and language, which together with the expansion of civilization have facilitated the growth of heteronomous controls – all of which is discussed in the next chapter.

CHAPTER NINE

HETERONOMY, MEMORIES, LANGUAGE AND CIVILIZATION

The top-down expansion of memory and the development of language could have featured in Chapter 3 under 'anatomical and physiological adaptations' for they must be linked to the enlargement of the hominin brain and changes in the larynx. I have held back the wider discussion until now, that is, until after the chapters on heteronomy and its consequences, so that the human dependence on memory and language can be more fully understood.

I wish to suggest that both the use of heteronomous controls and the subsequent need to find defences to those controls have been significantly enhanced by the expansion of two functions within the hominin brain. The first is an increased capacity to store memories with a long-term top-down recall, which allows those memories to be retained, manipulated, rethought and juxtaposed at will. The second is an increase in the use of sound (words) to convey meaning thereby creating a verbal language that can be used to transfer memories from one hominin mind to another. Over the last (say) fifty thousand years these two abilities have increased greatly, and they have facilitated all the projects associated with civilization and the social contract, including systems of law and order, as well as religious and social codes of behaviour.

9:1 Heteronomy, memory and the emotions:

Humans are able to retain, manipulate, rerun and juxtapose their memories in ways that are impossible for the bottom-up animal brain. Animals have less need to convert their short-term memories into longer-term memories because they do not fall victim to controllers that insist that the controlee remembers their heteronomous commands and directions.

To be able to function in a world of heteronomy the mind had to increase its memory capacity and develop the ability to access and manipulate its memories via top-down neural pathways. This enhanced ability has created a secondary priority system which meant that the hominins have developed two ways to access their memories and two ways to trigger behaviour, either bottom-up, or top-down.

For the animal, memories are linked to each of the five (or more) senses: sight, sound, smell, taste and touch. They are stored in the brain and recalled and used bottom-up as and when necessary. However, these memories are, in the main, locked into the individual animal's brain and cannot be recalled top-down, which means they cannot be manipulated in the mind or directly transferred into the mind of another animal. Thus, a sheep does not directly tell another sheep that it has eaten some lush grass that can be found over the hill; or a fox does not tell another fox where to find a goose to eat. Animals do observe the behaviour of others and they do follow the leader to a new feeding ground, understanding (non-verbally) that the leader is motivated by a memory of a worthwhile food source. But this is not one animal having direct access into another animal's brain as in the case of heteronomous control.

In coming to understand heteronomous control we need to recognize the predicament of the controlee. Heretofore, any danger that an animal faced was well established and predictable and each species had evolved bottom-up methods to deal with them (for example, programmes for maintaining a flight distance, taking cover, fleeing or fighting) that were set in motion autonomically. Once the hominin (controlee) was held and then brought under heteronomous control she/he had no bottom-up programmes capable of dealing with this threat. Yet the controlee strives to exist (*conatus*) which meant that she/he had to find ways to regulate this new, unsolicited, unwelcome relationship with the controller and the imposed content had to be organized in the controlee's mind in some new way. The question is, how is a relationship as complex as this regulated and maintained?

Bottom-up programmes are of little use in this situation because the requirements of a heteronomous relationship are unique, unpredictable and dependent upon the level of control that the controller may or may not enforce. This is where the ability to store and manipulate memories top-down becomes imperative and where no doubt a rigorous selection pressure came into play, for only those who could react top-down would have been able to survive the rigours of motor-control and the unprecedented and threatening demands of heteronomy.

The controlee must remember and act upon the controller's instructions, and construct and remember its own responses (defences) to every incident of motor-control (no two controllers are the same). These memories are even more essential when the controller issues delayed instructions, such as, "Don't do it now, wait until I tell you to start," "Meet me at sunset," "Do this in a week's time,", "Don't do that until the autumn." By harnessing the use of speech, heteronomy has (alas) become an exceedingly efficient method of control. Instructions can be planted in the controlee's brain which will be remembered (reinforced by fear) so that they can direct behaviour even though the controller is not present – "I was told to go to the river in the morning to fetch water," "He says I have to dig this ditch." No animal functions in this way. In addition, things not immediately present (even miles away), that cannot be seen, smelt, or pointed towards, can be brought into the realm of attention using words, such as, "Did you hear, yesterday a man was killed in a landslide in the Philippines?" Where is the animal that would 'say' anything like that?

This ability to open one's mind to the inputs of another and remember their top-down instruction (as well as the threat that creates the imperative of compliance) for weeks, years or even decades is unique to the hominins. From the perspective of the controlee (under the threat of punishment) it is essential to develop this ability, but seen from the perspective of the 'original (animal) order of things' it is a crippling weakness. Animals do not open their minds in this way, which guarantees that they do not act as a repository for the directions, wishes, needs, desires or untruths of other people; some of these thoughts may even have originated thousands of years ago, but have remained extant due solely to the top-down transfer of memories down succeeding generations.

Consider what the minds of the original autonomous protohominins had to accomplish once they fell under the thrall of heteronomous control; every instruction, wish or desire of the controller had to be memorized by the controlee so that they could be enacted and then the memory of that response had to be stored so that it could be used to rerun similar heteronomous behaviour in the future. Heteronomous life is lived via these memorized instructions, from simple things such as, "Don't forget to say please or thank you," to ways of cultivating the garden, to the rules of being part of a football team or a military unit or belonging to a secret society or simply of doing one's job. Once they have been placed in the controlee's mind they remain there, as memories, casting a shadow over the controlee's behaviour, often long after the controller has moved out of any direct sphere of influence, "I couldn't do it, my late grandmother would have been distraught". Animals are free of all such considerations.

Heteronomous behaviour (compared to autonomous behaviour) is always stilted because it originates from an external controller and is sourced from memories, which, by definition, are not a feature of the immediate present. It is the human reliance upon the top-down memories of heteronomous controls that creates the gulf between them and the spontaneous bottom-up behaviour of the animal.

In the animal and the hominin, conscious awareness (not to be confused with self-consciousness) plays a simple but crucial part in the mind's ability to fix and store memories. Consciousness is a heightened level of graded awareness that is triggered by the safety/danger scales in response to the actual or potential dangers to which the individual needs to pay attention. It results in wakefulness, awareness, alertness, and a readiness to react.

A function of conscious awareness is to register that an event is being witnessed or experienced, which in effect, marks it with a tag of 'personal remembrance' which then enables the individual to recognize the event as having occurred before (that is, it is 'known' as a past memory, embedded in time with a fixed date). This means that it can then be used, checked or, if necessary, rerun time and time again. If you do not remember that you had seen something in the first place, then it cannot be recalled and rerun as a memory because there would be no conscious awareness of its earlier presence.

An animal stores memory of such things as food, specific locations and other individuals not as words but in forms associated with the five (or more) senses, sight, smell, taste, touch and hearing, which are then accessed bottom-up. After motor-control, the hominins would have continued to remember things in these ways but in addition they developed a top-down system that is able to remember the controller's instructions plus the controlee's responses, including such things as, how did the controller respond last time, was the outcome satisfactory, where will it lead, what could I be made to do in the future?

Once these musings become memories that can be accessed top-down they are available to be run, rerun and juxtaposed and any consequences considered before they are enacted by the individual. These top-down deliberations are essential to regulate heteronomous life, as is the ability to access the outcome of the deliberations at later dates. Here is the essence of heteronomy, every response must be checked, double-checked and tested to see that it meets with the controller's requirements before it is enacted, this is a stilted process compared to the smooth, natural, self-governance of the animal.

Manipulation of thoughts (mimicking and outrunning the manipulations of the hand) allows for limitless conjectures, trials and reruns so that they can be juxtaposed and examined against each other. By using words and names this has become a remarkably fast and efficient process and it facilitated the breakout from the bottom-up programmes which, in turn, led to the explosion of human creativity, none of which would have appeared were it not for motor-control, anxiety and heteronomy.

Limited top-down mental organization will have started well before the development of spoken language. In the beginning, emotional states such as anxiety, fear or contentment were the means (as they are in the animal) by which the individual conveyed information around its body and mind, often in a form that can be observed by others. The behaviour of an autonomous animal has much less emotional content than that of a heteronomous human because, broadly, they are free of conflict and so lead more placid and contented lives. Some emotions such as shame, embarrassment and pride are uniquely human because they relate directly to the failure or success of heteronomous defences.

Emotional content has a priority which means that it is memorized and attended to, thus it is readily (autonomically) fixed in the mind to become a major internal pressure influencing future behaviour. It is an effective way to (non-verbally) influence (change or control) the behaviour of the controlee as she/he witnesses the anger, fear, rage (or contentment) of the controller because it generates in the controlees their own fear, anxiety, rage (or satisfaction) in response to the demands of the heteronomous controls.

Note there is always a sleight of hand inherent in the operation of heteronomous control regarding emotional content. The emotion is generated internally (autonomically) by the controlee and the controlee is programmed to act in accordance with the 'steerage' of the emotion that is generated by her/his own body (anger, fear, contentment, shame, horror). Thus, the emotional content of the controlee is self-generated even though the punishment (contempt, ridicule, violence) is inflicted by the controller. This means there is no need for the controller to give instructions as to 'how to be afraid' – it arises spontaneously from within the controlee and the controller(s) needs only to trigger this response to benefit accordingly.

The controller can be quite relaxed when he/she starts to say such things as, "You will be punished if you don't do it," but the fear and tension generated in the controlee, as the implications of the punishment are considered, rise to the point where compliance is assured. Thus, heteronomous information always enters the mind with an associated emotional content that is linked to the level of threat posed by the controller.

Obviously, the slave in chains will fear the slave trafficker and the victim (rapee) will be terrified by the rapist. At other levels of coercion, parents, teachers and others do not necessarily have to punish the child physically; the threat of punishment or rejection is usually sufficient to ensure compliance. These highly emotional states, associated with heteronomous control, are stored as memories in the mind and so they have a lasting (even permanent) effect which makes them highly significant, for they power the remarkably high levels of heteronomous activity that makes up perhaps seventy per cent (or more) of human behaviour. It is this emotional content (anxiety) seeking discharge that has powered the rise and development of language.

9:2 Heteronomy and language:

Horses, guinea pigs, otters, deer, shrews, elephants, giraffes, snakes, fish, birds and all other motile living forms that are capable of auditory communication are mute about the use of words. It is blatantly obvious that words are unnecessary for animal life and have been so for hundreds of millions of years. This is because no animal has the 'power' to become an intraspecific controller, nor suffers the indignity of becoming a controlee, so animals are unable to place heteronomous controls into the minds of conspecifics, which means they have no use for the deceptive power of words.

I suggest the motivation to 'name' is driven by the need to impose and react to heteronomous controls. This seems a more plausible explanation for speech than the current theory that claims it was a fortuitous result of a greater intelligence, which additionally carries the puzzle as to why no other living creature in the history of evolution has needed to develop the ability to speak.

Words are a tool of human control, they are the vehicle by which heteronomous controls are imposed and negotiated (and eventually understood). Spoken words (and written words) are an extremely efficient way of transferring memories, instructions and unsolicited information from one brain to another so they have played a key role in the rise and rise of heteronomy. Autonomy acts silently (pre-verbally), heteronomy acts by using threats, words and language to achieve its controls.

Originally most heteronomous controls were transmitted by means of growls, grunts, screams and cries, which eventually turned into words. The controlee's compliance and defences are marshalled in words, which mean that every time you use words, or think by means of words, then this is not the genuine autonomous bottom-up 'you' that is acting in this way. Words are how heteronomy is transmitted, without motor-control there is no heteronomy, without heteronomy there are no words. Words are not so much an advantage, they are a necessity in a world of motor-control.

Animals are alert to everything that goes on around them. They are alert to sights, sounds, smells, touch, thermoregulation but they are not alert to words. Speak to an animal (apart from heavily domesticated animals) and it will disregard everything you say. They live successfully using sight, odour, sound and taste and these are senses by which the early hominins experienced the first motor-controls, rapes and beatings. The victim's distress is experienced by means of the characteristic odour of sweat and blood, by observations of pain, emotion and sound. These sounds would have been grunts, cries, screams, squeals but initially no words.

Obviously, words have had a stunning uptake even though they require a prodigious memory and initially they are (for many) somewhat difficult to remember (and spell). The language explosion has facilitated writing, printing, radio, telephone, television, computers and the Internet, perhaps now that we can record virtually everything we will realise that none of this colossal human effort would have been required had we been able to maintain autonomy.

Animal memories are stored as visual images and/or olfactory, auditory, gustatory and tactile stimuli, which can be recalled when necessary by the bottom-up programmes. It is impossible to convey to another person the taste of what one tastes, the feel of what one feels, the vision of what one sees, the smell of what one smells because these sense receptors are unique to the individual and their stimulation cannot be manipulated in one mind and transferred to the mind of another by using the 'sense data' of that sense.

Only sound (converted to sight when written down but still reliant on the underlying sound) can be transferred to another's mind in a form that is guaranteed to be 'heard' (understood) in a universal way, thus sound (speech) has become the sense of choice for the implementation of heteronomous control.

We are the only animal to use a verbal language that has a spoken and written form. The experts say that some form of speech may have started to be used one hundred thousand years ago and by the time the hominins moved out of Africa seventy thousand years ago they had developed a useful language. However, I suggest that some enhanced vocal communication must have been used (and gradually expanded) from the moment of the first motor-control. The victim (controlee) would not have remained silent. In a paragraph associated with Cratylus (5 BC) it says, "In the heaving confusion of the perceptible world nothing is fixed, so thought can gain no foothold, and nothing can be said." One might add that nothing needs to be said because the thoughts of an individual animal have no need to gain a foothold within the bottom-up programmes that direct their behaviour. How else could we gain a foothold in (or break out from) the continuous flow of nature other than by holding it down (fixing it by means of our hands and motor-control)?

The flow of life has a built-in acceptance of events as they appear and then disappear into the past. Once this flow is halted, held or challenged by motor-control then a space opens in the hominin mind where memories can be held, re-examined and manipulated in an attempt to find top-down solutions to the intractable problems created by this violation of autonomy; words, names and speech have become pivotal shortcuts in this process.

In the primal control situation, the male holds the victim, he will growl and threaten her, she will scream, and he will force her to adjust her body posture to his desires. The male will develop rudimentary top-down thoughts in his mind regarding his intentions. The female (now exposed and vulnerable) will also develop top-down thoughts seeking to understand her plight and the creation of escape strategies. She will vary the volume, tone and pitch of her cries to try to convey as much agitation as possible, in the hope that it may deter her attacker and/or call for assistance.

The physical and psychological problems of being caught, held and penetrated (while anoestrous) will leave memories (scars) of unresolved behavioural responses in her mind which she would later rerun and juxtapose time and again, seeking a safer outcome. Even without full verbal language her emotional agitation will be visible to others and she would point, gesticulate, and scream to alert others to her predicament trying to construct alliances.

From this point on, as the controlee and the controller seek incompatible outcomes, different sounds would gain new meaning such as, 'Help,' 'No,' 'Away,' 'Stop,' 'Go,' 'Sorry,' 'Yes,' 'Good,' 'Bad,' 'Come here now,' 'No, don't go there.' Clearly it is the victims (controlees) that are in most need of the benefits that come from verbal communication as they utter the above verbal defences; later they may also say to their sisters, "Don't go there," "Get sticks or stones," "Keep away from Kurt," (who now has a name), or "Wait here where it is safe." Victims need to remember names and details (he came from over the hill) if they are to seek retribution, deliver justice, take avoiding action or simply tell their family. Whereas the controllers (the attackers), applying brute force, are not (at this early stage of development) likely to be particularly concerned with the victim's details (often it is better not to know the name or personality, just call them yobs or gooks) personal details may spawn sympathy, even give rise to empathy, a fatal flaw in those wishing to control others.

Eventually the technique of naming would spread to more ordinary situations and objects, words such as, 'tree', 'fruit', 'meat', 'drink', 'stream', 'sleep' would enter the vocabulary, information that primates seem to transmit with a combination of limited vocalization, body language and a form of telepathic understanding. This expansion of use is guaranteed because, without adequate defences, everything becomes of interest to the hominins in their desperate search for safety and this has led to the extraordinary fact that all objects, individuals, positions, movements, emotions, defences and dangers have been given a word or name. It is no longer a question of whether the object, or thing, is relevant to a species, with the hominins everything in the whole world (even the universe) is relevant because absolutely anything may be relevant to heteronomous controls or defences to those controls.

Names are neat snippets of information associated with their referent; they can be readily stored in the mind where they form a succinct (shorthand and short cut) mental dictionary that can be accessed at will. With the use of top-down neural pathways, names, be they fact, fiction, abstraction, from the past, present or future, can readily be rearranged mentally by the manipulation of memories and this loosens the close tie between the word and its referent. Once words can be rearranged mentally it is only a short time before virtually anything and everything can be rearranged physically. Houses can be built, rivers dammed, territory claimed, states created, mines dug, the role 'wife' invented, relatives disowned, atoms split, nothing is beyond consideration or introduction.

Presumably, animals think silently to themselves in the (pre-verbal) language of their senses, rerunning their last actions, 'stalk rabbit', 'catch rabbit', 'eat rabbit', 'drink in stream', however, nothing about this content is problematic, there are not any unresolved difficulties because everything is taken care of by the bottom-up safety/danger scales. Animals do not 'name' each other verbally because they are identified by means of an individual odour (but they cannot be controlled by means of an odour); their minds remain closed to any direct interference from others which means that they work out everything for themselves – that is the essence of autonomy.

The animal has safe intraspecific relationships, which enables it to 'rest contentedly', it has no need to say anything because first, it 'knows' what is going on, and second, it has no need to transmit that information to others because they also, via their own senses, know what is going on. Autonomous conspecifics do not enter top-down relationships, they simply maintain an honest acceptance of each other's presence, they do not get into another's mind to control their behaviour. The underlying animal system of communication is also present in the human for it is non-verbal communication that stops us bumping into each other. This can become quite sophisticated, for example, when driving into town along the main road I may see a car some way ahead that wishes to enter from a side road. I slow to create a space between my car and the car in front. If the space is of sufficient length the other driver understands my 'invitation' (is grateful and acknowledges that fact by raising a hand) and moves into the flow of traffic. There is meaningful communication here (in a dangerous situation) but not a word has been said. This non-verbal, voluntary (non-controlling) communication usually passes unregistered but it shows that we could (like the animal) get along quite well without speech if there were no heteronomous controls.

Speech is not restricted to communication between two (or more) people; much searching, thinking, mulling over and planning is carried out in the mind by silently talking to ourselves (about our innermost thoughts), indeed much of our brain time (perhaps fifty per cent) is taken up with these private personal monologues. This is necessary as every individual must work out her/his own strategies for dealing with the controls she/he experiences; these strategies are rehearsed silently in the mind before an acceptable version is made public (which illustrates just how vulnerable we are). An animal whose autonomy is secure has limited need for this type of mental activity (does the badger think to itself, "My spots are awful, shall I or shan't I go out today?" or "I am bored, my toys are broken, what can I do now?" or "I will go to the badger set down the road to see if one of them will come out for a walk"?) Next time you speak silently to yourself, closely attend to the mechanics of the practice for here is the manipulation of memories, "I said to him, he said to me," "Shall I, or shan't I?" "You go there, I'll stay here." Heteronomous control could not be imposed or enacted, or defences to controls constructed, without it resulting in this silent conversation in the mind; it is the manifestation of top-down manipulation.

Names have broken free from their referent, they can be spoken, or thought about, at any time depending on the whim of the speaker or thinker; 'rainbow', 'cellar', 'spleen' can be spoken or thought about, out of context with any present time experience of these objects. By breaking free from their referent, words have played a major part in improving defences to motor-control (including mankind's search for autonomy). However, severing the immediate link to the object has had other significant consequences, for it has become possible to construct lies, fantasies and false emotions, even false memories. Fiction may be used by controllers to further their plans, or it may be used by the controlees seeking to improve defences, by constructing scenarios in which the opponent is tricked, misinformed, deceived and placed at a disadvantage. There is virtually no limit to the fiction and lies that can be transmitted verbally (by comparison, truth is very limited in its range).

Speech that is contemporaneous with the event to which it refers is a comparatively rare phenomenon (for example, a commentary on a football match, or a horse race, or the Lord Mayors parade – or a boy saying, “Look at that fat woman,” or a mother calling her children to the dinner table). Mostly it is rerunning of thoughts or memories of past events or making conjectures or predictions about the future (for example, talking about ‘what did or did not place with the latest girl/boyfriend’, or ‘what the boss said’, or ‘how we nearly won the match’, or ‘memories of the delicious meal we ate last weekend’, or ‘what Jesus did at Galilee’, or ‘what will you do for me if I vote for you’, or ‘remember that holiday we had ten years ago’).

All of which highlights the fact that, at base, our use of speech is trying to resolve difficulties and problems, for we are ‘seeking answers’ or ‘discharging tension’ every time we speak. However, what we are seeking predates words, we seek autonomy and equanimity of mind and when that is achieved there will be no need for words, we could regain the muteness of the animals.

The comparison with animals is noteworthy; would a cow tell another cow that she was about to eat a tasty bunch of grass, or a snake announce that he was going to sleep in the sunshine? Would cats, if they were able, talk about what an Egyptian cat did three thousand years ago? Would horses talk about past winners of the Roman chariot races? Would the descendants of the elephants that went over the Alps with Hannibal trumpet their ancestor’s achievements? It all seems very unlikely, so why do we spend so much of our time doing the equivalent of just this? Why do we have such an interest in these things when animals have no concern whatsoever? Is this simply the bonus of intelligence and human exceptionalism, or is it part of our desperate need to seek status and find safety from motor-control?

Human languages must be manipulated top-down, they must be learnt and used in much the same way as any other heteronomous control and they always carry with them the clout, fear and uncertainty of motor-control as well as the uncertainty of incorrect use. Originally our ancestors were autonomous animals, they experienced no top-down heteronomous controls and they had no words in their minds. We evolved to become hominins struggling to deal with the trespass of heteronomous governance and our minds became full of thoughts structured in the form of words. The controls, compliance and resistance involved in these new heteronomous relationships had to be managed in some way and it is now clear that once they were converted into sounds (speech and words) then controls could be memorised, thought about and manipulated acoustically before being converted into the behaviour of self and others.

Words get their weight (clout) from their ability to gain the attention of the listener and they do this by their association with the punishments of motor-control. Consider the three words "John, come here." Remarkably, they have the power to trigger attention and demand a response, to the extent that an obedient John immediately complies with the meaning of these sounds.

Animals have calls that represent specific dangers, vervet monkeys, for example, have different cries that represent a snake, a hawk and a leopard, whereas humans have thousands of words that can be strung together in the multitude of ways required to deal with the myriad dangers that arise from the trespass of motor-control. It is not just John, but Chris, Gillian and Frank can be picked out and made to respond in different ways. People and objects are named so that they can be manipulated by others. Animals had never been named before the advent of the human, they did not ask to be named nor did they need to be named.

Animals do not use words because words are the messengers of motor-control and they are used to create the penetrative relationships by which heteronomous control is conducted. They are building blocks that covey the nuances of control; even words relating to freedom are predicated upon controls and earlier restrictions.

Words and their associated meaning are difficult to acquire; we are not born with an understanding of them (which points to the fact that we should treat them with extreme caution), and so they must be drummed into the mind before they are remembered. Learning to speak, read and write is part of the child's early experience of control, correction and compliance.

Words are spoken, spelt, manipulated, and written down in a considered way that meets exacting standards of spelling, grammar and meaning, and to be able to do this the speaker has to carefully direct the larynx, tongue and lips (and pen) in a precisely controlled way – this is not casual behaviour. The energy for this drive is not generated from standard bottom-up animal behaviour; instead it arises from a dire vulnerability to heteronomous governance. A small innocuous word obtains its clout simply because it is linked to the dangers of motor-control. It resides in your brain with the emotional charge associated with its heteronomous connection. Words are vicious little things because they come demanding attention at several levels, they have meaning, nuance and directions attached to them (as well as the complexities of spelling, pronunciation and grammatical use). What else in the animal world demands such exact top-down attention to detail? Not sex, not food, not the weather, not even the predator (for you just avoid or run away from a predator). It is true that smells and tastes do have exact olfactory and gustatory details, which trigger responses, but they are set in the bottom-up programmes, so they do not have to be learnt by top-down repetition and analysis for them to function and be memorized.

Words come as a package; you buy into a complete system that needs years of training imposed by parents and educators. The only way out of all this is to realise that no animal, insect or bird, in millions of years of evolutionary history, has ever thought or uttered a single word. They have no words in their head because words are completely superfluous to the successful function of an animal living in a world that is free of motor-control.

Words can be very powerful; they can articulate ideas, achieve compliance, violence, punishment and exclusion, they can be blasphemous, even treasonable and they can engender love and reconciliation. Words can be used for control or defence of those controls; you cannot have one without the other. When used within a constructed defence they can be a force for good and eventually they may help science and philosophy come to understand the human predicament. The paradox is that none of these words, not even those used as a force for good (improving the lot of the controlee) would be required if, in the first place, we had not fallen victim to motor-control.

9:3 Heteronomy, history and civilization:

Words under top-down memory recall are the key element in the creation of history and the projects that have given rise to the development of civilization. Before considering some of the implications of this it should be remembered that the cloud of heteronomy, under which we all live, is barely recognized, so the judgements we make regarding such things as family, society, morality, government, wealth and status are all made with little or no understanding of the fact that humans are the only animals to experience and become hostage to the intraspecific heteronomous controls which have created a cognitive dissonance that blocks out the equanimity of our underlying autonomous animal mind.

Having such a major element of human behaviour hidden in plain sight creates a serious problem, for it means we do not understand our true nature nor our natural place in the world, hence our unease.

The catastrophic loss of autonomy that followed the introduction of motor-control has driven the hominins to continually seek a safer existence and this has resulted in an inexorable search for truth, change, improvement, a better way of doing things (which hopefully will be maintained until autonomy, safety and equanimity are restored). Over time this has led to all the projects that have created such things as, countries, kingdoms, governments, dictatorships, religious groups, laws and regulations, which together have become known as 'civilization'.

Civilization is a constructed existence and without the top-down ability to store and recall memories all the information that is necessary to make history and create a nation would be lost; it would simply drift away into a non-reclaimable past. Animals with their bottom-up neural pathways do not have a history or civilization, they do not manipulate and recall their memories top-down hence they do not commemorate the birth of past matriarchs, or erect statues in memory of legendary males.

Civilized man is a very recent invention, say, twenty thousand years at most, an exceedingly short time in evolutionary history. So, although we defend our own nations, our current lifestyles and our religious and political groups with a fierce loyalty, the concepts they represent are no more than transient whims that can become a threat to our survival. Laws, taxes, duties and regulations of civilization are difficult and costly to maintain, and so they must be enforced by ever more burdensome restrictions, controls and punishments, driving us further and further away from the desired autonomous control.

In any discussions about history and civilization there are always two aspects of heteronomy that need to be considered – the justifications of the controllers, who mostly have precedence because of their power and dominance, and the lesser-heard pleas of the controlees who always have their autonomy overridden.

Controllers are adept at planting layers of justification to obfuscate the severity of the controls they inflict, for example, the rapist says, "She asked for it"; the slave owner says, "It's economic necessity"; the husband says, "She vowed obedience"; the head teacher says, "The child has to be educated"; the boss says, "You are paid to work here so you do as I say"; and the parent says, "Discipline is essential, I punished the child because she disobeyed me", the state says, "It is the subject's duty, we protect you and in return, you do what we demand." The controlees can do little other than complain, "This is too harsh," "Why me?" "I do not want to do it," "Leave me alone," or "It isn't fair." The outcome is a battle of 'justifications' between controllers and controlees which the controllers usually win and, to a large extent, this is the underlying dynamic of human society.

The journey towards civilization commenced the moment the first hominin was held beyond the point of submission. Motor-control creates such injustice that it inevitably gives rise, in the victim's mind, to the idea that the violation should not have occurred. This alternative scenario immediately gives rise to further alternative scenarios in which defensive improvements are made and safety restored. However, the individual needs to construct this 'safer' lifestyle from within and in contradiction to the continuous barrage of heteronomous controls that she/he is unable (does not have the power) to avoid. Does the badger say, "If only I didn't have to do this," or "I must find a way out of this terrible predicament"? We now have a situation where society inflicts motor-control on its own citizens as a means of establishing a social contract designed to protect good citizens from bad citizens.

If we continue in this way, then the total of 'control' (rather than the severity) imposed by the social contract will exceed the controls experienced before the contract. The task now for civilization is to reduce its reliance on motor-control and find a way to restore autonomy to all humanity. We need to absorb the fact that humans are not at the pinnacle, quite the reverse; they embody a tragic error, a devastating misstep of evolution – the 'fall' from the animal norm of autonomous governance. The climb out from this low point of heteronomy and control will be arduous; the controllers will try everything in their power to keep the controlees in a state of compliance, the solution will necessitate the nurturing of individuals that understand the beauty of autonomous governance.

Civilization is reliant upon language, however, language is Janus-faced, for not only does it enable the transfer of truth from one mind to another, it also facilitates the creation of fictions (lies and half-truths) that, with sufficient repetition, are turned into an 'accepted' reality (faux facts). Language is not subject to natural restraints; with language you can create gods, popes, heroes, royalty, generals, headmasters, mythical heavens, men on the moon, human exceptionalism, white heterosexual male supremacy, or any artificial rank or status that the controller desires – all are disconnected from natural reality but they are nevertheless used by humans to maintain the structure of civil society, for these constructions are indispensable to the enforcement of heteronomous controls. Animals have no need for any of them – can you imagine your dog getting into a position of power where it can say, "I am a general," or "I am the Emperor, from today all dogs will bow down to me"? Can you imagine a dog writing a book of canine history, or can you imagine your dog reading about dogs from the past, or dogs that should be revered or, more remarkably, a work of fiction about dogs that exist only in the imagination of the dog that is the author?

A hominin's ability to live in the world of heteronomy is largely dependent upon her/his ability to memorize the instructions and directions that are placed in the mind by others. Once memorized, these instructions can then be passed into the minds of the next generation as they too fall under heteronomous controls. As this process continues, down the generations, a historical record is created. Animals have no means to deliberately pass on the detailed information that is held in the memory of one individual into the memory of another, so their brains are not encumbered by the potentially limitless clutter that can create life-threatening inefficiencies and mental breakdown.

It is only the hominins that have had to accept this burden for it is the inevitable price of heteronomous control. Having never experienced motor-control the animal has no need to record and maintain history. This means that they do not hold opposing rationales in their mind that create the conflicts that spur them into battle on fear of death, capture, subjugation, rape or torture. In addition, without the ever-present background threat of motor-control, top-down discipline is unimaginable and unenforceable, so any hypothetical animal 'army' would not be able to muster a platoon because the participants would simply disperse (that is the strength of autonomy).

In warfare, humans engage in intraspecific top-down violence that has, in evolutionary terms, a very recent origin. With any violence that is organized top-down there is virtually no limit on the numbers that can become involved, given a powerful enough leader (or controller), millions can be made to participate in a focused, concerted, disciplined action. This excess is far removed from anything that can be generated by animals reliant on their bottom-up programmes and it illustrates the inherent (perhaps fatal) weakness of our dependence on top-down heteronomous control.

If we were able to curtail the control element of motor-control, humans would gradually return to living an autonomous animal life and the need for civilization would fall away – bears, despite their large teeth and claws, do not need courts, prisons, schools, governments, kings, queens, presidents or prime ministers. Whereas we so-called 'civilized' humans readily attack our conspecifics, we bomb defenceless mothers and children, we deny the poor equality, so they are left to live in squalor, we pick on minority groups, and we eagerly humiliate our children by sending them compulsorily to schools (where half are destined to become relative failures). We continually override the autonomy of our fellow human beings, yet we claim we are 'civilized' and to support that claim we wear clothes to hide our genitalia and refrain from urinating and defecating in public – a truly remarkable reversal of all animal behaviour and these latest twists have come about in less than a hundred thousand years.

The truth is, we have lost our autonomy and as a result we have become conflicted, anxious, ashamed, vulnerable, lost and we inflict heteronomous controls upon others without any thought or hint of shame. We have become detached from the natural order because we have lost the freedom of autonomous governance that is effortlessly experienced by every one of the animals every minute of the day. In its place we resort to controls, and create lies, fantasies and grandiose projects that are all doomed to failure because they are not based upon the stability of autonomous truth. Hence, we cause enormous damage, we harm ourselves and we degrade the environment that we should be sharing with all the other species.

What they think of us is not recorded, for as their habitat becomes damaged and polluted it diminishes around them, yet they continue to live in the present, relying on their bottom-up programmes that have (in the shorter term) no defence to the threats that humans create at breakneck speed. This is the logical outcome of heteronomous control: we end up fouling the planet, our own nest; it is a symptom of our pain (as with bed-wetting enuresis), the physical evidence is defiantly visible, yet the underlying distress is concealed and difficult to resolve.

Words, speech, language and civilized behaviour represent hazards in that they need to be taught by means of top-down instruction. Animals do not teach their young in this way because it is inherently too dangerous, yet humans have come to rely upon it – hence our troubled relationship with compulsory education, which is discussed in the next chapter.

CHAPTER TEN

HETERONOMY AND EDUCATION

The top-down development of spoken language has created an intractable problem for humankind; language is not acquired bottom-up, it must be taught top-down to every generation.

The written form, including spelling, is even more difficult to acquire hence the need for the intensive techniques of instruction and education of the young, which is unknown in any animal species. Originally this was restricted to the scribes, the wealthy or the very able but this limitation eventually came to be seen, as unfair. So, as a civilizing principle it was established that all children should be given access to education (taught to read, write and count as a basic minimum). To achieve this universal principle, it needed to be compulsory otherwise, for various reasons, many would not take part.

This allegedly well-meaning control has turned out to be a catastrophic miss-move because it has led to the dire situation where children are conditioned (throughout their formative years) to having their autonomy overridden by compulsory heteronomous controls. Not only is attendance compulsory but so too is the sitting of examinations to test ability.

After prolonged and bitter struggles, the treatment of women, black people, Jews, homosexuals, and various minority groups has improved in the last thirty years but the last great inequity, the compulsory confinement and testing of children in our schools has not been seriously addressed, indeed it is barely mentioned in discussions of equality, discrimination, emancipation and mental ill-health.

Children have no voice, they are locked down under the tight control of their parents, teachers and the state and very few adults are prepared to fight on behalf of their rights. Because of their age and size, children are unable to defend themselves physically, nor they can campaign and articulate their own needs; hence they are an easy target for bullying adults who subject them to scandalous degrees of coercion, compulsion and confinement. Parents control their children and believe that it is their 'duty' to train them to be obedient, indeed parents, except for the few who opt out under the Home Education Act, are under a legal obligation to the state to send their children to school; all at a time when the child should be experiencing a relatively carefree period of exploration of their environment, enabling them to express their autonomy, develop and mature before adulthood.

This situation is of interest in the context of this book because it represents a particularly blatant attack upon the child's autonomy, which we ignore seemingly because adults are completely insensitive to the pain and damage that children suffer. All of which forces the child into compulsorily absorbing an astonishingly large amount of heteronomously sourced information that the less retentive memory is not designed to accommodate, so for many of the pupils, stress and mental damage are an inevitable consequence.

This harm to the less able is considered acceptable collateral damage in the rise of the next generation of the academic elite. Retention of memories always has the potential to overwhelm the vitality of the individual, for unless one's inessential memories are cleared from the system, they become burdensome. The events and relationships that comprise the whole of human history are too numerous, too various and too complex in their entirety for any individual to absorb, understand and remember in a single lifetime, so even the best recall is staggeringly selective.

Extant individuals of any animal species are tasked with living out their lifespan in current time, information about self's own past experiences may be recalled from memory but they are relatively limited. But to compulsorily deny to a child what would have been its autonomously chosen present-time experiences (for example, playing outside with friends) so that her/his mind can be filled with past-time information, such as remembering people and events that existed long before the child was born, is a very bizarre outcome. Today, this has become even more bizarre, for computers now have far better memories than any human, yet children are constantly tested to see how much information they have retained in their mind.

Compulsory education involves control, obedience and loss of autonomy on a scale that adults would find intolerable; it is violence against the child, their life and work is regulated, and their freedom crushed. As this discussion proceeds focus your first thoughts on the experiences of those children that make up the lower fifty per cent on the academic scale, who become the 'failures' of the system. For in the main they do not (cannot) receive any compensatory payback for their suffering and confinement. Unable to achieve the necessary grades, they are denied access to higher-status tertiary education, they are less likely to find highly paid employment and are therefore unlikely to achieve a high social standing. So, the years of testing that this lower group is forced to endure is primarily for the benefit of the higher group, as they need a benchmark above which they can register their success.

It is inherent in this type of testing to have a fifty per cent failure rate, that is those below average on a bell curve distribution, and because average is well below elite, it is probably nearer to eighty per cent that become classed as failures (certainly by those at the top who have a good degree(s) from prestigious universities.) All of which means that the larger (lower) group are systematically relegated for the benefit of an elite minority.

At present the carefully orchestrated euphoria for the successes of the few masks any consideration or assessment of the damage that is inflicted on the majority. All of which means that individuals within this group spend the whole of their early years being constantly informed (overtly and covertly) of their inferiority (as were (are) slaves, women, black people, Jews and homosexuals), a practice inherent in all discrimination.

The comparison is telling, for in these cases there was a long period of denial by the abusers that the victims (controlees) had any rights that could be abused, it was thought that they were either stupid, or simply enacting their allotted roles in society, or had themselves to blame for being treated harshly. At the time it was 'obvious' that women needed to obey their husband and black people needed to be owned by others to curtail their insolence and work harder, just as it is clear today that the less able schoolchild does not deserve to get on. Today, it is thought that children are unruly, undisciplined and ignorant hence they need to be told how to behave, what to wear, what to think, what to say (in an 'approved' accent), what they should learn and be repeatedly told that they must work hard to show they are employable.

Children are certainly too young to have their autonomous behaviour respected, so it is in their best interests (allegedly) to place them under a regime of compulsory education (from an increasingly early age) until they are eighteen because the state (capitalist, socialist or communist) needs an educated workforce. Today we increasingly hear that our children are not sufficiently well educated to compete in a global market. The state has this degree of control over our children's lives because it wields a power of 'inclusion or exclusion' that stems from the ability to motor-control. This will continue until humans learn to respect each other by re-establishing and maintaining an ability to live autonomously.

For adults, compulsion is treated with the greatest caution and its imposition must be justified, supervised by law and restricted to the shortest possible term. To fall under orders of compulsion adults must either break the law, be directed by a police officer, be the subject of health quarantine, be sectioned under the Mental Health Act or be conscripted in wartime. Only in these exceptional circumstances is compulsion accepted and then only because it is ensuring the safety of others.

One may argue, for example, that if obese people and addicts were compulsorily made to address their problems it would be highly advantageous to both the individual and the state. They are not made to do so because they do not represent a sufficient threat; enforcement would create insurmountable difficulties and it would be morally wrong to do so. However, this is not the case with children, because they are small, and they do not understand their own plight; to the shame of all adults, children have no rights in this area of compulsory education. They are forced to attend school with little or no thought about the moral legitimacy of its mandatory underpinning, let alone the psychological damage it does to the child. Not until the child, new to school, gets up to "go home to see my Mummy" does she/he find that the school gate is locked from the inside and that she/he has been tricked and confined. I suggest that it would be intolerable, illegal and physically impossible for any larger, stronger adult group to be treated this way in peacetime.

This is a shocking situation so how are these violations perpetrated?

The government forces the child to be compulsorily educated with the threat of legally enforceable punishments on the parents. This means that it forces the parents (by law) to do the distasteful, morally indefensible work, which may involve dragging a refusing child (one's own son or daughter) into the school premises where she/he is confined. Which raises another question of a dubious legality: it is the child who is compulsorily detained yet children as young as this cannot be legally responsible for their actions; hence the parents face the fines or imprisonment if the child tries to assert her/his right to freedom of movement; surely a blatant deceit is being perpetrated here.

However, it is not just that every child is now under a compulsion to attend, they then find that they have entered a regime where they are regularly assessed on their ability to pass tests, the results of which stay with them for the rest of their life. It sets up a desire to succeed in some and an expectation of failure in others; is this really of benefit for the whole of society and should society organize such a pervasive system of inclusion and exclusion?

By the age of five, half of our children have been classed as below average (that is, they have less retentive minds than the other half) and most of these children keep this classification for life. One needs to be reminded that once you test a whole group it is certain that fifty per cent must fall into the category of below average and be classed as academically less adept. And one also needs to be reminded that these so-called 'failures' are human beings with as much right to have a fulfilled life as those who are above average. Elitist testing has the deliberate objective of exposing the (so-called) failings of others; here are the tyrannical roots of all discriminatory systems, which are designed to select winners and reject losers.

Do the benefits for society that accrue from promoting the skills and success of those with retentive minds really outweigh the permanent damage that is done to the half with less retentive memories? That is, is ignoring the distress and harm that is done to the losers a legitimate, intelligent or moral response to the situation? Those with retentive memories would still retain more information than average in a non-compulsory system of education.

The human mind is subject to variation between individuals. When considering the retentiveness of memory there is a natural continuum that ranges from those with exceptionally retentive minds to those who do not have retentive minds. Given that all the individuals on this continuum are human beings, and all will live out their allotted lifespan, it is indefensible (incomprehensible) to compulsorily select (expose at a young age) the bottom half of this continuum and label them as inferior to the others. This is a gross abuse of adult power, it is tyrannical, and it is behaviour that is not seen in any other animal species. Worse still, the memory that is being favoured here is an ability to remember heteronomous information. This type of memory is non-existent in the mind of the animal because it is the antithesis of functioning autonomously.

There needs to be a return to a desired modesty regarding the possession of a retentive memory (or any other ability bestowed by nature). Animals do not flaunt their mental capabilities even though individuals are likely to differ markedly. They certainly do not measure, record, tabulate and publicise the different levels of retention of which their minds are capable. These differences may result in significant benefit, for example, scarce food may be found more readily, innovative sites for shelter may be chosen, but these successes will not be recorded on certificates for others to see. "I found the most worms in a year," said the robin as it collected the prize.

This could be called secular blasphemy, for these differences (when they are evolutionarily meaningful) are 'recorded' by nature as variations in the survival rates of progeny which can only be measured long after the participants have died, so the difference cannot possibly be exploited as status or used deliberately for advantage in the lifetime of the original individuals.

So why is the degree of retentiveness of memory, measured

and treasured by humans, not subject to this truth? Simply because it can be exploited for short-term gain (and rejection) in our heteronomous world where one is constantly seeking defences to coercion.

If you could tell a badger that it was a success due to the retentiveness of its memory it would respond with a look of incomprehension and walk on. In evolutionary terms (in a world without motor-control) simply being alive is the only success that has any weight. However, in a world of motor-control, compulsory education has become a way of deliberately gaining advantage over one's conspecifics, but it comes at a terrible price.

It is said that in England in 2013 one in five children have a school phobia and fines for truancy have risen by one third. These problems arise from bullying by other pupils, unsympathetic or rebarbative teachers, or simply from a biologically reasonable fear of strangers and compulsory confinement. Have blood and hair tests been taken to see just how stressed a child is when it enters a new school, when it is disciplined, when it takes an examination, when it is bullied, when it misspells or miscalculates, or when it fails examinations? Every school should publish these results as part of their 'duty of care'; perhaps they would then come to understand the consequences of their actions.

Testing to ascertain how much of the heteronomous instruction has been remembered is a double assault on the controlee, which in the case of failure can lead to a triple assault in which corrections have to be put in place. People vary; these variations should not be understood as symptoms that require correction, assistance or improvement, for without testing and censorious assessments they would barely be noticed. The choice of which trait that is, or is not, chosen to be tested, compared and deemed desirable, or undesirable, is of interest.

Some children are better proportioned, better coordinated, have better skin tone, have better hearing or clarity of diction, better sight, thicker hair, or are more daring than others. It is not seen fit (indeed it is improper) that children should be examined (let alone compulsorily examined) and compared for these traits so that some individuals can be listed as being superior to others.

So, in what way is a compulsory examination to find the retentiveness of one's memory (or intelligence) different to these other examples? Is there any difference, or is educational elitism one of the most widespread forms of discrimination that has ever been practised?

If a new teaching method (or memory pill) were to be developed and standards miraculously improved, to the extent that every student passed all exams with the highest grade, the system would collapse, there would be no academic elite, and in those threatening circumstances the old elite would simply raise the bar to reinstate the fifty per cent failure/pass rate.

So, this is not a system, like the driving test, that is designed so that virtually everyone can pass, this is a system of inclusion and exclusion where at least half have to fail for the others to be seen to succeed. Shouldn't the children, who over a fourteen-year period are deliberately being selected and designated as the 'failures', be told of this fact rather than being led to believe that failure is their own fault for not being bright enough or not working hard enough?

The argument that compulsory education helps some of the poor or less advantaged to become educated is true, but it is a red herring in this argument, for the problem is not which individuals fail or pass but that half of the school population must be classed as relative failures for the others to be seen to succeed. So, even if a few students do improve performance and obtain higher grades it only means that someone else must be relegated to the 'below average' category, that is the fact of average. Competition has been taken to the point of cruelty.

Animals compete for mates or scarce resources, but the adults do not set out, in this determined way, to identify the young conspecifics that outperform the others and then favour their progress in the world, nor do they set out to identify the young that perform less ably and then discriminate against them. Even predators do not test the ability of their offspring to catch prey, this skill vital to the survival of the young predator is learnt instinctively (autonomically).

The lesson for educators in this simple example is that it invalidates every reason for human compulsory, pedagogical instruction and examination. There is certain to be a marked difference in the abilities of young inexperienced predators to make a kill, but the adults have no role in testing and tabulating this difference, nor in instructing the best techniques (other than by example). So, in the natural autonomous world, even with the skills of predation, which are so critical to survival, the adults do not instruct, test or select the skills of the young. All other animals protect their young while at the same time allowing them the freedom to make their own decisions (express autonomy); it is to be hoped that, one day, this will not prove to be as far beyond the ability of the educational system as it is today.

However, the most telling objection to the compulsory element of education is that it facilitates mental rape, the violation of the young child's mind. The child under compulsion, for many hours a day, must attend to the teacher's words, even if the child has, for deeply embedded emotions, quite reasonably taken a dislike to, or has a justified unease with that individual teacher. That is, the child has to allow the heteronomously sourced instructions, facts or commands of the 'alien' teacher into his/her mind; then, on fear of the punishment of failure, has to remember this information (retain it within what should be her/his personal autonomous mind) and then use this information in ways that are deemed correct by the teacher (school or state) certainly for the rest of her/his school life (and, as is the aim of indoctrination, probably for the rest of his/her life).

If judged by the standards of freedom set by nature for all other animals, this is aberrant (and abhorrent) behaviour and it is at the extreme end of the completely insensitive heteronomous controls of which humans are capable. Those in education, with their higher-than-average intellectual intelligence, lack the emotional intelligence to understand this glaringly obvious fact, whereas the children who used to played truant (when it was possible in the days before surveillance cameras) knew it in their bones.

Those who enforce the compulsory element of education are selected for their insensitivity to heteronomous control; they efficiently absorb and pass on information, but in the main they are not troubled by the loss of autonomy suffered by their pupils. The subject 'philosophy and history of education' no longer has the high priority in teacher training colleges that it used to; presumably this is because any alternative to compulsion is considered undesirable.

Animals do not trade knowledge (they may be granted deference when mature because of the knowledge and wisdom that comes with age), but they do not flaunt knowledge, neither do they use knowledge for power, pleasure, recreation or blatant advantage. Animals do not have voluntary lessons, let alone compulsory lessons from their elders, they learn autonomously using their own eyes, witnessing everyday situations, and that has always been the case.

There is not an animal on the planet that goes to school; there is no animal that is taught in the sense of being deliberately instructed. This is a useful indication (a silent lesson from nature) of the attitude to knowledge that humans should adopt; it should be reserved for private and personal information only.

This is because all heteronomous instruction has a serious inherent weakness, which is, how does the controller know what is truly in the best interest of another individual at any time? How can the controller know the internal state of the controlee's mind into which the heteronomous information is forced?

The damage that arises from this situation is disregarded by humans, yet it is of sufficient seriousness to be rigorously avoided by all other animals. Does the most intelligent baboon set 'pass or fail' tests for the young of her troupe and if she did what would she do with that information? Would she promote some at the expense of others, and if she did, would that be in the best interests of the whole troupe? Tell a badger that under compulsion it must go to a selected sett, put on a uniform, study studiously for many years and then sit examinations to see that it has learnt a sufficiency of information to be fit to be an adult badger.

Again, there is a quite shocking lack of respect of (or effrontery to) nature (God) in all of this, it is as if the child has not inherited the benefits of millions of years of evolution and is not perfectly equipped to absorb all the information that it needs using its own eyes and senses to be able to live for the relatively short period of time that is the human lifespan. By what authority do the heteronomous controllers claim their right to direct the behaviour of others? Of course, they have no right; compulsory education is simply an expression of the gross adult power that arises from the aggressive use of their hands.

Compulsory education is clearly the product of motor-control, for animals never behave in these ways. When a group of hedgehog's forage on your lawn do they designate which one of them is the brightest? Undoubtedly one is, but it is not a fact they choose to ascertain and display; animals exhibit an indifference, a modesty, about their own abilities that we would be well advised to adopt. It is motor-control (not intelligence) that allows heteronomous information to be forced into the minds of our children so that winners can be selected, promoted, favoured and exhibited, and losers rejected and discarded. The high tension that is always present in any debate about education is a sure sign that we do not fully understand its function and are anxious about its underlying structure.

Compulsory education is a relatively new phenomenon; first mooted in 1642 it did not become established until 1842 in Massachusetts, USA. Children between the ages of eight and fourteen had to attend school for twelve weeks a year. On average it is now thirty-eight weeks (190 days) with additional homework for the evening, weekends and holidays plus the tyranny of examinations. In addition, the age range has increased by eight years to include four to eighteen-year-old children (that is, virtually all of childhood); these are major increases, all in less than 200 years. Are we certain this experiment is of benefit for the whole society in the longer term?

One of the functions of the compulsory system is to inculcate obedience into the pupils, for it is a necessary requirement for the discipline of the modern workplace and compliance to the regulations of the state. It is the cornerstone of the move that has finally squeezed the last drops of free autonomous human 'being' out of all of us and turned us into heteronomous slaves indentured to fifty years of waged labour.

It was realised at an early stage that unless education was made compulsory many children would simply not attend. Universal attendance was thought desirable for several reasons, chiefly because non-attenders would fall behind and become disadvantaged. They would not be able to fulfil their 'duty' to a modern state, which required compliant citizens able to work in the new industries, fight in modern wars and pay their taxes. The rights of children were non-existent, or simply not considered, so it became a trap for the child and an unseen moral trap for the adults into which they all readily fell.

Obviously, the child is the primary victim, but the adults are also ensnared because they like to think they are doing good, helping the child get on, ("It is not too bad, school is only for a short time," they say). The problem with all of this is that none of it addresses the question that was present from the outset – it is simply immoral to lay out compulsory plans for other human beings and is especially wrong if the controlees are children, so it is undoubtedly an issue that at some stage will have to be addressed by society unless we are destined to succumb to harsher and harsher heteronomous controls in a dystopian world. One may ask why the majority participates when it is only a minority that benefit. The answer of course is that they would withdraw if they could, that is why it has been made compulsory.

Obedience and disobedience (with its inevitable punishments) are unique to humans, they are the classic indicators that at some stage a combination of power, grasping hands, motor-control, discipline and punishment have been used to violate autonomy.

So, one may ask, why is this taking place?

Virtually every parent remembers the pain and anxiety they experienced in their own school years and every parent knows the pain and anxiety that their children experience during their compulsory attendance at school, but everyone is mute. Meanwhile the pressures of examinations inexorably increase year upon year, impervious to the rise in childhood anxiety, depression and suicide rates (see mental health statistics, Young Minds 2016) for we must keep up with the Japanese, Chinese or any other country that appears to threaten our fragile self-esteem and our commercial competitiveness, (do English badgers fear competition from Japanese badgers?).

So, the child is confined to the school premises where it is subject to strict standards, timetables, uniforms and examinations imposed by adults for reasons that it is too young to understand. This is in keeping with the sadistic violence and abuse that has always been part of compulsory education, the beatings (corporal punishment was not finally banned for children until about twenty years ago), bullying, fear of ridicule, fear of failure, expulsion, punishments, detention, (for as little as an 'incorrect' hairstyle, an 'incorrect' skirt length or an 'incorrect' logo on a shoe) all of which has to be accepted by the children basically because (like torture victims) they are denied any protection or legal representation. Are children's-rights solicitors appointed to every board of school governors?

Pupils suffer heteronomous control, heaped upon control, upon control; we need to admit that the compulsory element of education is a form of persecution. It is certainly child abuse for it is deliberately planned and organized; in what other context can we find the equivalent of a child being compulsorily detained simply because she/he has reached the age of four, or sent to boarding school, away from family, at the age of eight?

The venomous bias of elitist education is present even in death, for if a young person dies, or is killed in an accident, it is always reported as more tragic or a greater loss if she/he was a straight A-grade student. (Do we hear that her death was a tragic loss because she struggled with mathematics or he was dyslexic?) So, why is this discrimination permissible when discrimination between white and black, or Christian and Muslim students, is rightly deemed intolerable?

Clearly, it is because the lower-grade students have no means of defending themselves, and so they are picked on by the elitist bullies. But why is this allowed to happen? It cannot be good or beneficial for the whole of society to first designate part of a normal distribution of abilities as inferior and then select those who exhibit this so-called 'lower' trait so that their presence can be publicized and contrasted with those who exhibit a superior ability (also part of the normal distribution), but this is precisely how discrimination in education operates.

All of this is listed because it illustrates the worst excesses of heteronomy, which are only made possible because we have hands, for when the child objects and runs away (takes the autonomous action of avoidance) she/he is caught by adults and returned. This inability to avoid is at the heart of all human problems. Once the child is caught and held then she/he may experience anything from rape, torture, punishment, or compulsory education, depending on the desires (whims) of those who possess the power to control; animals are fortunate not to be subject to such dangers.

Any grand talk of the compulsory element of education being 'in the child's best interests' is as ridiculous as saying that confinement in a zoo is for the 'benefit' of the chimpanzee. Similarly, ridiculous, are the current claims that teachers, faced with inattentiveness, rudeness, violence and even murderous aggression, need to inflict more discipline rather than less. Here we see the inability of the educational establishment to step back and to see how outrageous the level of the heteronomous control they inflict has become in the last 150 years. We have developed a shocking insensitivity to the wishes, desires, needs and experiences of other human beings, particularly our children.

With our hands we have gained the ability to confine and control others, now we need to gain the sensitivity to see the havoc it wreaks. It is not just that we instruct and educate our children, we test them to destruction. Academic elitism, like any other power structure (wealth, race, gender, royalty), is dismissively hostile to the group they subjugate because superiority is reliant upon (manufactured out of) their power to artificially create an inferior class, it is not a natural division. However, the facts of the alleged inferiority are of little concern (indeed the differences at stake can never be a valid reason for discrimination); the elite group may claim their status and entitlement by birth, skin colour, gender, race or religious faith or, in the case of academic elitism, by the superiority of possessing a retentive memory, but their claims are always predicated on their ability to wield the power to include and exclude.

The sheer folly of this elitist approach is highlighted by the fact that it is not wisdom. We now have the most numerous and highly educated generation of adults in the entire history of the world who have presided over, witnessed, caused and done nothing about the pollution of the planet and the steep decline of animal species that is now taking place. We have overseen the most astonishing rise in world population levels, indeed, according to the author and biologist Christian Schwägerl, humans and their farm and domestic animals now account for ninety-seven per cent of the larger animal biomass in the world leaving just three per cent for the larger wild animals. We have witnessed and done nothing to stop the shocking degradation of the environment, soil erosion, debris in the sea, forest reduction, fresh water depletion and global warming, which have all come about because of our frenzied overconsumption. At the same time, we have also presided over a staggering rise in the incidence of human depression, anxiety and drug dependency (recreational and medical). All of which is a fitting testament to the fact that academic excellence has no inherent connection to wisdom, or survival. Compulsory education has played a major part in this downfall for it creates high levels of tension that have to be discharged and this has given rise to levels of consumption and non-essential activity not seen in any other species, to the extent that humans are damaging the planet faster than it can recover. (Some claim that the size of the human population is evidence of the fact that we are the most successful species that has ever lived. However, a population of this size means that we have broken through the natural restraints that limit animal populations and in so doing we have inevitably started down a path that will lead to a catastrophic crash.)

A small child, for those who need to be reminded, is nature's gift of a perfect new human being and has a complete set of inherited abilities necessary for survival in a world without motor-control. But today in a world of motor-control a child can be corrected, humiliated and classed as a failure depending on as little as (say) the exact position of an apostrophe.

This pedantic insistence on establishing 'correctness', above the hurt and discouragement of the child, reflects the dangerous controlling principles extolled by the compulsory educators as they exhibit their own vulnerabilities. Having plans for others, in the form of a curriculum which is deliberately detailed and invasive (you must learn this by the age of five, and be competent in this by the age of ten) is a sickness of the adult mind for the pupil's present-time well-being becomes a secondary consideration to the primacy of the educators. This long-term pernicious interference in the behaviour of the young is a human construct, it is not found anywhere else in nature.

Eventually, compulsory education will be one of the great disgraces of humanity on a par with slavery and the control of women by men. Indeed, its net reaches far wider than slavery for virtually all the children in the world are now subject to the coercion and restraint of compulsory education, all pitted against each other, studying harder and harder to keep ahead, happy to consign fellow students to the lower categories, thereby ensuring their own success.

This brutal competition is organized by adults, governments, educators and administrators, which highlights the degree to which a frightening insensitivity rules our lives. There is a point where heteronomous instruction simply overwhelms autonomy and the individual's struggle to maintain personal integrity. Many pupils in the educational system are now reaching this breaking point (suicide, depression, anxiety, anorexia, self-harm) but the victims are disregarded, their plight becomes classed as inevitable, unfortunate or unavoidable, they have become collateral damage. The compulsory element of education is a sin.

The joy of seeing a child autonomously exploring its environment and learning instinctively has, like a rare butterfly, been lost from these shores. It will not be seen again unless we hear and heed the daily cries of anguish that emanate from every school in the land, cries that are integral to the acts of compliance, heteronomy, obedience, failure and punishment, which are demanded by an elitist educational system that is compulsory.

All heteronomous controls and the individual's response to these controls merge in the human mind to create a 'personality'. The personality enables each one of us to bargain and negotiate our way through life when we are unable to establish autonomous governance. Personality is the subject of the next chapter.

Footnote: For clarity, (and for those who have succumbed to apoplexy), my criticism chiefly focuses on the unforgivable compulsory element of education. In a heteronomous world a genuinely voluntary system of education comprising short periods of instruction may be tolerable. But, as the animals preach in their (unheard) sermons – any individual living a fully autonomous life has no need whatsoever for any heteronomous instruction.

CHAPTER ELEVEN

THE HUMAN PERSONALITY

Heteronomous controls have become so pervasive that humans are unable to understand (have no feeling for) the 'mental stance' that results from leading a truly autonomous life. The only clues we have come from animals and by looking closely at their mental attitude to life we can start to see just what has befallen our species.

The first thing to note is that non-human animals have existed and evolved over a much longer period than the hominins and they have done so without any recourse to heteronomous controls or top-down mental activity. This is not because humans have superior abilities, better brains, or the ability to use a spoken language (which animals lack), it is because humans must deal with the extremely difficult problem of coercive motor-control that animals have been fortunate to avoid.

Every individual has an understandable, justifiable dread of being held against their will because such vulnerability readily leads to injury and acts of unsolicited penetration of the mind and body. This dread is exacerbated by the realization that any assault by a powerful controller is capable of being repeated, virtually at the will of the controller, and as these violations cannot be avoided, the victim builds an 'expectation' that she/he will have to endure and comply with every demand and violation. This is so extraordinary, it is almost beyond belief, but it is the result of the selection pressures that have driven hominin evolution. We have become somewhat like the sufferers of 'Stockholm Syndrome' (where captives fully identify with their captors) for under heteronomous control we have become vulnerable and so compliant and complicit with the wishes and directions of our controllers that we 'choose' to look upon many of them benevolently.

We are at a disadvantage from the moment we are born; we lack adequate defences, so we are hyperalert, looking, watching, and fearing any form of control. This is not letting events unfold as the animal does, humans must work out the best way to behave to survive and defend themselves from control, which means that we have a stance to ourselves and to the outside world that is anxious and troubled. There is never any dispute as to who has governance of the animal mind, it is always the autonomous individual who never becomes victim to the heteronomous control of any conspecific because it can always turn its back, walk away and escape. In contrast, the human mind is unable to protect itself and for most of the time falls under heteronomous governance.

At the interface of this 'battle for governance' between the controlee and the controller is a top-down 'buffer zone' (the template of compliance) that has evolved to allow the controlee to absorb the heteronomous demands of the controller and adopt personal techniques for dealing with the trespass and then to enact them as if they were her/his own even though they are (initially) the directions of the controller. The human personality develops within this buffer zone, hence the animal, facing no intraspecific threat of motor-control, does not (by definition) have a personality – only a character.

Animals do not negotiate with other members of their own species as to how they should lead their own life. Whereas the moment the human becomes victim and is motor-controlled, she/he is forced into a relationship with the controller and this gives rise to a unique and intense focus upon self in a way that simply never occurs in the mind of an autonomous animal.

The first moments of reflection for, say, a victim of rape, abduction, or forced labour, would be bewilderment, "Oh help, oh help, how did that happen to me, why me?" The controlee, reflecting on survival, would realise with horror and anxiety that she/he was not only defenceless to the present control but also defenceless to future controls and abuse. This reflection creates a focus in the mind around which a new core of thoughts emerge that concentrate on the deficiencies (inadequacies) of self, "I have been attacked," "I have survived the attack," "I am vulnerable," "I cannot escape," "What can I do now?"

This is the start of 'self' organizing a continuous stream of top-down thoughts that stems from the awareness that if 'I' am going to survive, 'I' need to create my own defences, for 'I' can no longer rely (as in the past) on defences being provided, bottom-up, by nature. Once self is required to deal with this degree of difficulty it is bound to focus on itself in a unique way and it is from the repetition of these urgent, individual, top-down searches that the concept of 'I' emerges.

The self is seeking the right, correct or successful mode of behaviour – "I do not know what to do," "I will try to deal with it this way," and "If it doesn't work I will have to try another way," "It is essential that 'I' find some way to comply with these controls and then, if possible, find ways to avoid or reduce them in the future."

There are two systems of governance, bottom-up and top-down, but the top-down system is open-ended in the sense that it is constantly changing. The bottom-up system is under the tight regulation of evolution and changes can only become established via the evolutionary (slow) route of the survival of progeny, whereas top-down organization of the mind is able to make instant changes so that it can react and behave in any new way the controller demands. In addition, top-down organization is (when necessary) capable of a rapid expansion; often, many controllers (some new) must be accommodated within a short period, each with their different (contradictory or complex) priorities, rewards and punishments, which must all be registered, memorized, enacted and integrated by the controlee.

It is this proliferation of memories and compliances that introduces such complexity into the human mind and creates the multi-facets of the individual's personality. The top-down self, your 'I', your personality, is made up of other people's thoughts, wishes, desires, instructions and your responses to them. They have been placed (forced by means of trespass) into your mind, mostly in the form of words and emotions so that they trigger your actions in ways desired by the controller. ("Today, Class Five, open your book at page eighty-nine.") You are unable to erase these heteronomous controls and empty your mind because the behavioural patterns they generate are fixed by means of threats and emotions that are very difficult (perhaps impossible) to ignore.

Hence an individual human is a hybrid, an amalgam of those controls that have entered the mind and remain lodged there (plus residual pieces of her/his bottom-up self). There is no way (apart from enlightenment) to clear them out, or erase them, because they have become your top-down personality, the person you now are. Your personality is impure because your mind is full of other people's inputs; this is the nature of being in thrall to heteronomy. Your ego, your 'I', your 'self', is a hotchpotch of other people's controls. If you live in Japan you speak Japanese, if you live in Russia you speak Russian. If you are Church of England, Catholic, Methodist, Muslim, Jew, Hindu, Buddhist, Conservative, Socialist, Fascist, it is very likely that your parent(s) were also minded that way. These divisions are not divisions seen in nature, they are contrived, constructed and maintained (as words) in people's heads in the form of heteronomous ties. A horse is a horse is a horse wherever it is in the world, irrespective of its parents or those it meets.

If it was possible to strip out all heteronomous inputs from your mind so that it returned to its original autonomous condition (as it would be in a world without motor-control) then our minds would operate in a way like that of the animals, with no heteronomous thoughts directing our actions and interfering with our autonomous governance. We would eat, sleep, defecate, urinate, pass wind, yawn, do nothing, explore, have sex (very occasionally), give birth and rear young (if female) and then die, that is it, nothing more. Any significant change comes about because of selection pressures operating on timescales way outside of an individual's lifespan, so the animal has no top-down concerns about improving its own behaviour.

Here is the paradox: animals that are autonomous do not have a name, an ego, a self, a personality or an 'I', their minds remain uncontaminated and pure. Humans, whose minds are made up of other people's inputs and controls, have a name, an ego, a self, a personality and an 'I' and their minds are ambiguous and equivocal.

Humans are proud of this situation and will often defend (to the point of death) the heteronomous principles that have been forced into their mind by others. Here is the underlying falseness of the grandeur, status, exceptionalism of the human, it is unnatural, it is planted by others and it creates an aggressive heteronomous brashness that obliterates the natural, autonomous, confidence of 'being', leaving the vulnerable human mind full other people's wishes, desires and instructions. As soon as you are forced to internalize other people's heteronomous outputs then your top-down self-conscious mind is made up of information and behavioural norms that have been placed in your mind by means of trespass.

The question then arises, who are you?

If this heteronomous person is not the true you, then where are 'you'? What would you have been like without these heteronomous inputs? You do not know for you have never experienced your true self (the basic, autonomous you) because it is unavailable once a top-down mode of thinking is in operation.

Your personality is largely made up of inputs that originated in a time and space away from your own mind and body, some even before you were born, in your father or mother's childhood. Indeed, the origin of many of the controls to which one is subject is often unknown to the controller let alone the controlee, which makes it more bizarre that it is the controlee that is held to be responsible for the heteronomous controls she/he must enact. The chain of responsibility is even more complex than this since the controller (such as parent, teacher, spouse) also feels a responsibility for the way in which their instructions are enacted by the controlee, hence they monitor the performance and issue correction because they also fear criticism and rejection.

If heteronomous controls were to disappear, a different state of being would emerge in the human mind and enlightenment would dawn. The default position of the animal mind is emptiness, whilst the default position of the human mind is to be permanently full of other people's inputs (full of the 'garbage' of heteronomy, it is not even one's own garbage) – that is the measure of how far we have fallen. The underlying dynamics of personality are predicated upon how one is controlled by others, how one complies with others' directions, and how one controls others, that is, whether one is rejected or accepted, excluded or included. It is the relative success, or failure of these responses that goes to make up (is expressed as) the personality. Thus, the personality is moulded by several factors, first, the objectives of the controller(s) and the type and severity of control they enforce. Second, the way in which the individual can deal with these controls, such as compliance, collaboration or resistance, and third, the ability of the controlee to construct successful cooperative alliances with family, friends or fellow controlees. And fourthly, with a remarkable reversal of roles, the way in which self learns to motor-control others to improve her/his defences (while remaining a controlee in other relationships).

All of this generates an unsettling and unstable background in the mind to which self must adjust. The personality emerges because autonomy has been compromised and a struggle ensues to balance the conflicting interests. It can be seen from these conflicts that autonomy now has three modes of expression:

1. True autonomy (full, pure, bottom-up autonomy) enjoyed by all animals. It suffers no heteronomous controls and has *no* ability to control conspecifics, so it wields *no* heteronomous power. Humans never experience full autonomy.
2. A trespassing autonomy, or rogue autonomy where (due to the ability to motor-control) the human controller now has the top-down power to impose heteronomous controls on conspecifics. The trespassing controller usually benefits from these controls whereas the controlee is placed at a disadvantage. The trespassing controller fails to respect the autonomy of others and becomes the dominant element in the joint behaviour (relationship) that is heteronomous control. Most (if not all) humans become trespassing controllers at some stage in their life, some more than others.
3. A compromised autonomy where the full autonomy of the controlee is overridden by the top-down heteronomous controls and governance of the controller. The controlee submits, in fear of punishment, to the controller's wishes, desires, and instructions and thereby becomes the compliant element in the heteronomous control. Virtually all humans suffer a compromised autonomy from their early childhood.

Human existence is a mix of being a compromised controlee and a trespassing controller and the personality reflects the balance that the individual strikes in coming to terms with this duality. Thus, if you avoid all controls and refrain from controlling others your mind becomes like that of an animal, 'empty of personality'; it simply allows the bottom-up programmes to act out their protective behaviours autonomously and autonomically and in so doing displays the true biological character of the individual.

Your 'I', your personality, is nothing other than a means by which you deal with your manipulations and it is constructed out of the interplay of individual physical inheritance with the restraints placed on it by parents, and others, and the restraints placed on self by self.

Consider now, if 'I' were to 'give up' this construct of personality, because it is mainly of external (heteronomous) origin planted by parents and others (and because animals live without it) – how would 'I' justify myself, how would 'I' defend my principles, how would 'I' negotiate contacts with the opposite sex, how would 'I' measure my worth, how would 'I' keep tabs on the vast amount of information that comes from society?

That is the point; animals do not need to keep a register of these things because the only reason to do so is to be able to defend oneself from motor-control. So, the personality is born out of motor-control and it exists solely to negotiate external governance. The original 'animal me' did not have a name, it led a life of 'being in the world' experiencing the simplicity of 'hereness', yet I am now incapable of simply 'being here' because I am defending, planning, devising, originating, organizing, inventing, formulating, lying and negotiating my own existence and I must do this because others had (and have) the power to force their heteronomous controls into my mind.

Animals do not have a personality, or an active sense of 'I', because all their actions 'happen to them' bottom-up, that is, they are not planned or organized by their conscious mind, top-down, ahead of the activity. The animal does not associate the triggering of its bottom-up programmes with the top-down sense that 'I have worked out what to do and will now put it into operation'. It does not have a platform in its mind that is sufficiently differentiated from the bottom-up system to support a top-down structure of thought that could give rise to the concept of self and 'I' through which it could negotiate relationships. The animal does not experience heteronomous controls, so it has no need for this type of defence, it does not take ownership, credit or responsibility for the bottom-up behaviour it enacts, and this surely is the only honest response because the animal's basic behaviour was selected by evolution many millions of years before the present individual was born.

The human sense of 'I' promotes itself as 'being in charge' of the individual when, at best, it is in charge of an out-of-control situation continuously trying to balance the heteronomous demands of others. Therefore, the personality as such is a complex, indeterminate mental creation. The element within the human personality that facilitates heteronomy is treacherous, for it is at odds with its basic autonomous self. Other elements try to preserve a limited autonomy as best they can, but they are not very successful given our vulnerability to motor-control and heteronomy.

Self 'knows' it has inadequate defences, that it is anxious and that it does not comprehend how it found itself in this situation, so it is not surprising that it is unable to understand its own role or function in the world, for its most pressing task is to comply satisfactorily to the next tranche of heteronomous demands that are forced into its mind.

The human mind, from an early age, responds to mental contact from others and is able to make mental contact with others. This penetrative mental interplay has developed into human social intercourse to an unprecedented degree. Evolution has selected the child that is able (born) to be receptive given that it is bound to experience (cannot avoid) the controls and heteronomous directions of others. The child is primed to respond, it does not know about the ultimate horror of motor-control (rape, slavery, abduction) it just responds because that increases its chances of longer-term survival.

Neither does it know the scale of the problems it will ultimately have to face having allowed its mind to be penetrated in this way (just as the female is at first unaware of the problems that will arise from her permanent receptivity). Being receptive and responsive opens the human mind to virtually uncontested penetration (especially in the early years), which may reduce the initial violence of coercive control, but it creates a mind focused upon compliance rather than the free expression of autonomy. Responses to the many different demands imposed by others gradually build up to become a 'repertoire of responses' that form the basis of a personality. If these responses are formed in a loving, caring environment then minimum harm is done to the emerging self. If the child is forced to 'respond' to physical, sexual or emotional abuse then lasting (permanent) damage can occur because the victim understands and experiences the virtually unthinkable fact that she/he is defenceless and has no escape from a repeated loss of autonomy. All of which creates severe mental stress for the child (or adult) for it now must construct behaviour that satisfies its own minimum requirements, whilst at the same time having to comply with harmful heteronomous demands over which it has virtually no control. In extreme situations insanity is a 'sane' way to preserve the last vestiges of autonomy that are vital for existence (see Chapter Twelve).

11:1 Animals use no words:

The use of words and language were discussed in Chapter Nine but there is an additional question that is of interest here. Wild animals do not use or respond to any words, so their minds and their thoughts are organized in a totally non-verbal way. If the animal mind does not use any words, how does it function without them?

It is very difficult to understand, or imagine, what the human mind would be like (or have in it) if it were empty of words (and speech). The first thing to note is that, although there is not a single word in the animal mind, all animals function perfectly without them. Humans have a top-down ability to manipulate a myriad of words and they have become the vehicle by which we can readily recall the past, live in the present and make conjectures about the future.

A *New Scientist* article, 18th November 2006, is of interest here for it reported that, "a man recovering from a stroke that had virtually abolished his capacity for speech, including self-talk, described the condition of total wordlessness as being like confinement to a continuous present."

This seems to be somewhat like the animals; with no top-down ability and no verbal ability, they are reliant solely on their ability to experience the world via their senses so everything, for the animal, is (and has always been) predominantly in the present. This suggests that the ability to search and manipulate the past and present is only required by a species, like ours, that is struggling with the trespass of motor-control. Without top-down manipulation, all memories would lie in the brain waiting to be triggered by the appropriate bottom-up programmes.

This is the animal condition and if we were able to return to a state of reliance on bottom-up programmes we would no longer be preoccupied with yesterday's, last month's, or last year's events (let alone those that took place 200 or 2,000 years ago, or events that may, or may not, take place in the future). Lectures, books, computers, television programmes and films all recall and rerun a multitude of memories to gain new insights, knowledge, or status; all are irrelevant to the animal who lives successfully in the present without them (does your cat watch the latest movie or read Shakespeare, Gibbon or Machiavelli?).

Try to imagine your own mind functioning without words, without a single thought being expressed in the form of words. All that would remain would be thoughts that are framed in the information of the senses such as, tastes, smells, touch, vision and spatial awareness. There would be other thoughts (awareness's) relating to bodily needs such as hunger, tiredness, waste elimination and sexual reproduction. And there would be thoughts related to memories that are triggered bottom-up. But none of these thoughts would be formed as words or speech. Whereas the human mind is occupied, either talking silently to self, or to others, and these verbal conversations take place throughout a large proportion of the waking day.

This is very unusual behaviour; trillions of other living creatures over 500 million years have not had a single word in any one of their brains, until *Homo sapiens* who, late on the scene, now have their heads crammed full of words (even senses, such as taste, smell and pain are converted into words, so they can be talked about). Try to estimate the time that the human mind is occupied with words throughout the day, perhaps it is seventy to eighty per cent; that is time occupied doing something that no animal does.

This absence of words creates a relative 'emptiness' in the animal mind. Normal animal thinking, in the sense of awareness of the senses, takes up some of this time, but there is still a considerable amount of waking time that is spent by the animal where it is alert, but its mind is less than fully occupied, which can be surmised by observing animals in their natural habitat.

The period when the animal mind is at rest is clearly not an anxious time, I would go further and suggest that it is a tranquil time, where the mind is simply maintaining a state of 'being'. If this is so, and if it is possible for the human to regain this state (once safe from motor-control) then the implications for human behaviour are beyond measure.

11:2 What is *not* in the animal mind:

Words and verbal communications are not found in the animal mind, nor are any behavioural patterns associated with heteronomy, so to quantify the relative emptiness of the animal mind we can start by concentrating on those thoughts and activities that fill the human mind but are not found in the animal. Listing things that are *not* found in the animal mind is rather a roundabout way of proceeding but it has the advantage of highlighting a major difference between the human and the animal that, like everything else, has been turned on its head by human exceptionalism.

The animal mind mainly registers non-verbal sounds, movement and odours; there is no top-down concept of 'I' that manipulates their experience of the world. There are no verbal heteronomous commands, instructions, orders, requests, desires or wishes that have been placed there by others and none that are waiting to be issued.

In the animal mind there is no feeling of responsibility that must comply with any top-down demands of others or to plan defences, or to do anything that has to be organized top-down by self or others (indeed the animal does not have the neural pathways to behave in a top-down mode). There are no worries that permission must be sought, or advice solicited, "Can I do this, or go there?" The animal either goes, or it does not, based upon its own assessments.

There is no wish to be stronger, younger, prettier, more handsome or more talented. That is, there is no wish to be anything other than the natural, present-time, autonomous self.

There are no top-down thoughts such as, "This has to be learnt and remembered," for stimuli that are of relevance to the species are stored and remembered autonomically by the bottom-up programmes.

For the female animal there is no subconscious thought in her mind that she cannot single-handedly defend herself from the male grasp and has therefore lost the ability to establish anoestrus. Thus, there are no thoughts that she needs to take great care (exhibit modesty) and form defensive alliances so as not to be molested by the males. This is not the case with the human female, for having lost the ability to establish anoestrus, she realises her defences are totally inadequate.

There are no thoughts in the animal mind that it must be cautious not to break religious codes.

No thoughts that the animal will get drunk or 'high' or will listen to music or watch television, or entertain friends.

No thoughts that the animal will organize a holiday, or will play sport or go to the gym.

No thoughts that the animal will discuss politics, philosophy, economics, or religion.

No thoughts of having to find a job, or having to work to earn money, or having to pay the bills, the mortgage, tax or insurance.

No thoughts of preparing a meal for the family, or having to do the washing, ironing, housework, cleaning the car or mowing the lawn.

These are just a selection of the everyday thoughts and tasks that fill the human mind. None of which are in the animal mind, so if none of this entered the human mind, in any form whatsoever, what then would then be left in the human mind? Almost nothing.

Thus, the empty mind would be carried by (float upon) the bottom-up programmes, it would simply register presence and observe the interaction between self and the external environment, thus enabling dangers to be spotted, and memories to be laid down for future reference (all done autonomically). That is, external events and one's own bottom-up behaviour are registered as occurring but the necessity for top-down mental activity to control and modify behaviour would be non-existent.

To clarify, animals have no thoughts other than those necessary to enact the next unit of behaviour selected by the individual's bottom-up programmes and these thoughts are structured around sound, odour, taste and spatial awareness, but not words. They will exploit difference if the need so to do is built into their bottom-up programmes, but they do not seek out, create or exploit difference top-down and their mind remains free of this type of surveillance.

So, when an animal lies in a pasture looking out over the valley, I suggest its mind registers presence and continuity by recording events and, when necessary, turning them into memories which, if prompted bottom-up, can then be recalled. It looks out watching for activity so that it can spot predators (or prey) and registers the movements and positions of conspecifics. All of which allows the animal to maintain a continuity of existence as its bottom-up programmes change their priority depending on the current internal and external circumstances. This means that the conscious mind of the animal (whether it is young, old, sick, healthy, male or female) is confident that the bottom-up programmes are equipped with the best possible responses necessary for survival – there are differences of character between animals, but these are not differences of personality for these only arise because of heteronomy and defences to motor-control.

11:3 Human collaboration, compliance and cooperation:

As the hominin brain developed its top-down mode of organization it allowed the controlee to liaise and negotiate with the controller to modify the (alien) heteronomous instructions. The area of the human mind tasked with this difficult role has now come to be seen as the 'personality' for it fronts the decision-making process and has the final say as to the way in which the individual enacts the controller's desires, wishes or instructions.

The personality is an essential element in the complex relationship of compliance and it creates a situation where the top-down mind, because it is designed to collaborate with the occupying force, can no longer be entirely trusted by its own autonomous, bottom-up (animal) system of governance – hence our unease with ourselves. The personality is primed (has been 'groomed' from childhood) to cooperate, collaborate and comply because it must deal with all the directions that penetrate its mind. This means it is on the defensive from the start; heteronomous intrusions are (in the main) not solicited, they are alien and unpredictable and because they are penetrations into the mind, they are always antagonistic to autonomy. Under duress, the self will collaborate with the controller(s) even to the extent of enacting behaviour that is damaging to self, having made a calculation as to the balance of harm. This is damaging but it can get worse: the controlee could perhaps comply mechanically, like a zombie, but the controller demands more than this, he/she demands an active participation and it is this active element, against self's better judgement, or will, (in which self becomes even more false or untenable) that leads to breakdowns in mental health. (See Chapter 12.)

Heteronomous behaviour has various elements ranging from willing cooperation to compliance under duress, where the cost to the controlee varies depending upon the personalities, objectives, and relationships of both the controller and the controlee, but it always falls short of behaviour undertaken autonomously.

Collaboration with a dominant controller is a complex process, there is usually an implicit desire of the controller to make the controlee do something that she/he initially did (does) not want to do: "You need to grow up," or "It will do you good," or "You will have to go, it is rude not to." The controlee is pressured into being compliant but in so doing suffers a serious loss of integrity because her/his own behaviour is essentially out of her/his autonomous control.

Personality is formed out of heteronomous relationships and personality has a role in the imposition of controls, for some of us are much more disposed to impose controls than others. Animals do not enter penetrative mental relationships with each other or with their offspring; they provide milk, shelter, warmth, protection but they do not control them or seek compliance. When a sow lies down the piglet can walk up to its mother's face, they will communicate by means of grunts and squeals but at no time does the sow penetrate the piglet's mind to instruct it, "You are to suckle teat number six," or "Lie down and go to sleep," or "Come here, line up, I want to inspect you before we go out for a walk." It is hard for humans to understand but all intraspecific animal relationships function smoothly without verbal contacts, heteronomous controls or the nuances of personality.

11:4 The vulnerability of the human personality. The uncertain self:

If we return to the author(s) of Genesis sitting in a village over 2,500 years ago thinking out the ideas contained in the phrase, "Who told you that you were naked?" the thought process must have been something like this – none of the animals in the village, or the surrounds, are clothed, none appear concerned by their nakedness and they do not appear to be anything other than the way they have evolved within nature. Why then did our human ancestors feel shame to such an extent that they had an urgent need to cover their genitalia?

Today, we have become so obsessed with clothes and fashion that the underlying problem is barely considered, but it remains as perplexing as it was the day Genesis was written. To have feelings of unease about a part of the body that has an essential function and cannot be changed is a curious phenomenon. No animal feels shame, modesty or anxiety regarding its natural essential genitalia. So how is it possible for an individual to set up a critical position against itself that in effect wishes its own body not to be seen?

Consider the hours that modern humans are preoccupied with their grooming, their hair, make-up, clothes, fashion, cosmetic surgery, and the reconstruction of body parts – it has become a major element in the activities of the day. Ultimately, all this dissatisfaction with the natural (as given) body arises from our vulnerability to motor-control and the difficulties arising from permanent receptivity. The basic principles of mammalian sexuality need to be understood before the human preoccupation with sexuality can itself be understood, and before the concept of female emancipation can be understood at its deepest level.

"Then their eyes were opened, and they discovered they were naked."

Here is a moment in the transition from protohuman to human. But how did they know they were naked, why were they embarrassed about their nakedness, why did they feel that it was wrong? This moment of awareness comes about when the female suddenly becomes aware that her genitals are seen by the attacker in a context that is quite different to the way she views herself (that is she available to be raped).

This knowledge of another context arises from her vulnerability to motor-control. The protohominin female living in the bottom-up animal world would not have thought that she could be caught and coercively mated when she was anoestrus. The protohominin male would not have thought that he could catch and coercively mate an anoestrus female. Nor would they, in a different situation, have thought that they could become enslaved (neither does the giraffe 'think' that it could be caged in a zoo); this type of information will only slowly dawn on the victim after the event (the child takes many years to fully understand why the paedophile abused her – even in adulthood it remains a puzzle).

Eventually the victim may come to realize (admit the fact) that what took place did take place, that she/he had been an unwilling participant and that in some way this has been accepted, acknowledged and assimilated by the self as it becomes part of self's undesirable, incomprehensible history of unsolicited heteronomous control. This delay in understanding essential facts about social behaviour has bedevilled human relationships throughout history and it arises because the controlee is never in full possession of all the facts because they are concealed in the controller's mind.

When motor-control entered the world, events unfolded that heretofore had not occurred throughout the millions of years of previous evolution. These new events then became selection pressures out of which new responses evolved, but in the case of motor-control the bottom-up programmes could not find an effective way to deal with the situation in a way that left autonomy undamaged. This, in turn, left the controlee seeking guidance top-down (from self, others, teachers or God): "I have been caught, I have been raped what can I do?" But neither the victim nor nature knew at that stage exactly what rape was or how it violated autonomy.

Or, in a different context, if I am being held in shackles, what can I do? Again, neither the victim nor nature knew what the essence of enslavement was. It has always taken a great effort for the plight of the victims of these traumatic events to be heard and understood. Even today it seems that a large section of the male population fails to understand the full implications of rape from the victim's perspective and almost the whole of the world's population fails to understand the gravity of the heteronomous controls that have become such a large part of human behaviour.

All Eve knew was that something terrible had occurred and that she would be somewhat safer if she covered her nakedness. However, if you could ask a mare, "Who told you that you were naked?" she would look blankly at you. She does not feel that she should be anything other than herself; she is not troubled by her own self, or by her own nakedness. Yet Eve, who by now had developed a top-down 'alternative self', felt she needed better protection, she needed to be clothed; her natural, naked, biological self made her uneasy, for her sexuality made her a danger to herself. Unable to rectify the problem, she developed a subtle ploy; she became 'ashamed' of her own nudity and this prompted the cover-up of her exterior body to reduce (a little) the extent of her violation.

She was able to justify the wearing of the loincloth (which originally must have been a very odd practice) because it did afford some actual (and symbolic) protection. Presumably the males went along with this because, by now, they had become half of a pair bond and clothing enhances the privacy and exclusion of that bond.

Once the original autonomous self had become differentiated by top-down control it created an inadequate 'personal self' that was able to be modified and improved, as and when necessary, by creating, manipulating and presenting different facets of itself to the world.

There is a nude self and a more desirable (safer) clothed self. God was upset when he saw that Eve was clothed because it was an incontrovertible sign that she felt ashamed; he was certainly not upset earlier when he saw that she was naked (and felt no shame). For the modern civilized mind to understand the full impact of this moment, imagine going to a field to find that, overnight, the horses had decided to wear trousers. The utter improbability, the unnecessariness, the sheer folly and stupidity, would immediately be apparent – indeed it would be laughable, yet this is what the hominins chose to do because they were under the thrall of heteronomous control and the female was unable to establish anoestrus.

Shame is defined as a humiliating feeling in one's own eyes, or another's eyes, because of one's own offensive, disgraceful or disrespectful actions (or those of an associate). It is felt in situations of embarrassment, dishonour, disgrace, inadequacy, humiliation or chagrin. Shame is fear of what people will say, of later retribution, ostracism or ridicule. No wild animal feels shame because they are, always free from heteronomous interference. Yet we also see here the 'expressive' element of shame, the human is visibly trying to say something through the expression of shame – the individual is struggling to find a safer stance.

Imagine being born into a world without motor-control, where nobody can tell you what, or what not, to do, where nobody can get into your mind, or your body – it would be total freedom, safe from any penetration.

Imagine an existence in which nobody can tell you that you are wrong, nobody can tell you that you are at fault, nobody can tell you that you have made a mistake, and nobody can tell you to correct your errors. Yet it goes even deeper than this, it is not just others who say these things, self has learnt to 'say' them to itself (and to others), "I should not be seen naked," "I must work harder," "Oh, I must have looked so stupid," "I wish people liked me," "I hate it, but I do it for my husband."

These are the thoughts and tribulations of the heteronomous human personality that cannot defend autonomy. From a very early age interaction with your parents, relatives, carers and teachers creates your personality; traces of their heteronomous demands and your responses remain in the mind as a framework for dealing with the next demands. The child is 'trained' via heteronomous control, to "Do this," "Watch for that," "Don't do that," "Be like that," "Learn this," or "I will hit you, punish you, be disappointed in you, or reject you." It is not possible (at present) to survive in the human world without the rules, codes, manners and training that result from this onslaught of heteronomous controls. In being compliant and obedient we develop a personality that is moulded by the controls that the individual experiences.

Personality is the interface between the controller and the controlee; a zone of acceptance and adjustment, an area of heteronomous chatter, bargaining and negotiating submission, compliance, collaboration and cooperation. It is a unique zone within nature in which top-down humans relate to their fellow beings. However, if the personality is constructed out of the heteronomous controls experienced by the individual then it can also be deconstructed when those controls become too intense and intrusive. This is the breakdown that takes place in mental ill health, which is discussed next.

CHAPTER TWELVE

HETERONOMY AND MENTAL DISTRESS

It is virtually impossible for a wild animal living autonomously in its natural environment to succumb to mental illness. The brain of the animal may be subjected to physical injury, birth trauma, seizures, tumours, poisons or parasites which will compromise its function but apart from those that are captured, tethered, domesticated, trained or confined in zoos (that is, come under the thrall of humans), animals do not become mentally ill.

All animals maintain a self-contained mental integrity because they are invulnerable to motor-control; their minds cannot be penetrated because no animal can hold another animal for a long enough period to insert the necessary heteronomous controls. This, in effect, means that there is an unseen barrier around the animal mind, which gives it a priceless, lifelong protection to all forms of heteronomy – a protection that was lost to the hominins when they fell victim to motor-control.

The hominins (as a group) have evolved out of a context of intraspecific conflict, where both male and female (especially the female) fall victim to incessant heteronomous controls that are a gross violation of their autonomy. The facts and issues regarding this situation are inherent to the human condition but they have not been disclosed, let alone understood or resolved. So, with this level of difficulty being played out beneath the radar, it is inevitable that humankind is sick. Humans exhibit a universal neurosis in their daily behaviour, and so they become susceptible to distress, malfunction and mental illness.

The less autonomous the behaviour becomes, the greater is the risk that it is liable to breakdown. This is certainly the case when heteronomous control involves harsh physical, sexual and or emotional abuse as these are areas in which the individual is programmed to defend itself with the utmost vigour.

Even in less harmful everyday situations, being able to enact the coercive controls of others requires remarkable mental agility on the part of the controlee. Compliance is always within parameters that are set by the controller and in the final analysis the controlee not only has to cede control of self to the controller but has also to understand precisely what the controller wants, on fear of punishment.

Remember, the controlee's mind is primarily charged to deal with its own basic needs for survival but this primary function (originally the sole function) now must be accommodated alongside the dire inequities and unrighteousness of the heteronomous occupation of the controlee's own mind. It is out of this implausible, disastrous and tragic relationship that the conditions necessary for the breakdown of mental health are created.

The top-down 'I' (personality) is the focal-point through which humans view and marshal their lives. The problem is that it is mainly constructed out of heteronomous controls, so it is not genuinely solid or strong (innate) like the character of an animal. The symptoms of mental ill health are a sign that the present sense of 'I' is untenable and falling apart.

In terms of mental organization, it is an extremely complex

problem having to tolerate being controlled by others and then adjusting to the difficulties of enacting their instructions. When coercion and conformity overwhelm autonomy, the controlee's tension levels become intolerable and if there is no effective way to openly express and reduce that tension then the pressure builds in the controlee's mind where (beyond frustration and anger) it converts the conflicts into neurosis and psychosis, which allows it to be dealt with or discharged in ways that are outside normal family behaviour.

The blight of mental ill health commenced the moment the protohominin female was coercively mated (raped) and she found that she was no longer able to defend anoestrus and autonomy. If she survived to become a mother her heightened stress levels would axiomatically give rise to deleterious emotional effects in the child who, inevitably, would experience either overprotection or neglect because of this poor maternal care.

This is certainly not the textbook 'good enough' parenting seen time and time again in the behaviour of wild animals. The first stressed female, together with her sisters who were in a similar predicament, created a new type of 'dysfunctional family group' that gave rise to the hominins.

These individuals were not autonomous animal-protohominins but hominins susceptible to coercive mating and heteronomous control. All of which meant that they became ever more watchful, neurotic, lost and frightened. Some females would fall victim to serious physical injury, having been forced to engage with the males well past the point of submission, but all would be mentally damaged by having their autonomy overridden and being subjected to heteronomous controls.

Also, they would necessarily rear offspring with a different, less contented demeanour; they may even have grouped into rogue bands where their difference (agitation, insanity) may bizarrely have gone some way to protect them from others. One thing is clear: these individuals are constantly uneasy, they are on the lookout, pursuing tangential activities to release their anxiety, and they would have set themselves apart from animal norms. For unless intraspecific relationships are conducted (more or less) harmoniously within the boundaries of autonomy there is 'nowhere safe to lay one's head'.

Any understanding of mental illness needs to be seen against this background. The mentally ill are conspicuous because they have suffered the worst excesses of heteronomous control (imposed against their will); at some stage in their history, their autonomy has not been respected, they have been unable to establish their own behaviour patterns, as victims they have not been 'listened to' or allowed freedom of movement and expression and in these troubling circumstances they have had to use the lever of 'difficult behaviour' to seek respite, change, or attention (even via depression) as a means of basic survival.

The rest of us (more or less) conform to society's norms and find our security in acting out conventional behaviour in the belief that we are successful, or respected, or hard-working, or intelligent, or gifted, or wealthy, or knowledgeable, or strong, or special, or loved, but these are symptoms of the 'madness of humanity' (or the 'madness of normalcy'), for animals do not need to justify themselves in any of these ways.

As soon as the controlee's mind is penetrated by the top-down controls of the controller, a neural pathway is opened through which others can subsequently enter. This means the controlee then has a dire vulnerability that will compromise personal integrity for the rest of her/his life. Once the controllers can implant their thoughts and controls into the mind of the controlee and make demands such as, "Do as I tell you," "Do that," "Don't do that," "Come here," "You must do it," then peace, contentment equanimity, privacy, security and full autonomy have been lost forever – one's life is no longer one's own. These are the facts of human normalcy, but they are unknown in the animal world.

Children may be subject to widely differing levels of coercion, discipline and punishments (which in effect is compulsory because the child cannot opt out) so, for humans, it is a lottery into which type of family one is born. When a child is motor-controlled she/he is forced into a heteronomous relationship with the controller, who is usually a parent or close relative. If the controls are applied with a light touch, that is if they are clearly stated, reasonable, not contradictory or too demanding, and noncompliance is not punished too severely then, remarkably, the top-down mind can adjust and comply (more or less) with most of the directions. The child has a chance of establishing good (or good enough) relationships with its parents and siblings and in so doing develops a personality stable enough to interrelate heteronomously; for humans this is normalcy.

However, for some, it can get worse, it is not just that your mind can easily be taken over and altered, it can also be damaged (sometimes irreparably) by the actions of aggressive insensitive controllers – "You have to marry the violent old man – you have been sold to him, he will take you away in the morning," or "I hated your father, I never wanted you, I wanted a boy, you have always been in the way, I wish you were dead," or "Come here, you have to pleasure me," (meaning "I am going to violate you"), or "Come here, I'm going to stub out my cigarette on your arm," or "You have to have a clitorectomy or the men won't like you," "You have to study harder, all our family have good degrees," or simply, "You need a good beating."

These are major violations and if the individual is subjected to a serious or prolonged loss of autonomy a shocking spiral of descent may occur which attracts others to inflict their own cruelties, unless the victim (patient) is fortunate enough to find sanctuary. In circumstances such as these it is useful to 'lose' your mind, so that your mind may be deemed not to be your own mind, or you are said to be 'out of your mind'. For, if you are not mentally 'at home', a buffer is created that reduces (somewhat) the patient's mental 'involvement' in the intolerable acts into which she/he is forced.

It is also possible that those who have access to the child's mind to demand total compliance are themselves mentally unstable, violent (even psychotic) or, equally damaging, that the child is at the behest of controllers who issue contradictory directives.

Mum may not be a loving mum because she is a woman who has experienced abuse herself and is seriously mentally ill, and or dad may be violent, drunk, disturbed or a sexual predator. In such circumstances the child would need to perform incredible feats of mental gymnastics to be able to understand and closely relate to such individuals. In fact, it is likely to be impossible, yet the child (unless protected) is forced to remain in close contact with such a parent often with no guidance or assistance.

In these circumstances the child has little option but to resort to aggressive behaviour and/or fall into mental ill health, attempting to merge truth with untruth and deal with the intolerable, illogical, heteronomous controls that have penetrated and overwhelmed its mind (and soul). In a situation where the adult dislikes the child, or suspects he is not the father, or when a child is thought to be gay or bad or evil, or when the child is used by the parents as a vehicle to explore their own abuse in childhood (to see what happens when someone else is severely beaten or sexually assaulted), then the child will be faced with problems it is very unlikely to be able to manage. The origin and purpose of the adult's control is hostile from the outset, it is outside of anything the child could understand, or with which it would wish to cooperate, it is psychologically damaging, and it threatens autonomy at every level, yet the victim can do no other than endure the controls and comply.

No animal falls victim in any of these ways.

Harsh compulsory control means that the actual inner

autonomous child is ignored. It can then be treated as a sexual object, or a toy or a puppet, mentally enslaved for the controller to play with, as if she/he felt no pain, or had no autonomy to violate, a situation that can only come about if the controller is without empathy, callous, hateful, stupid, and cruel. It is not possible for a child to deal with this onslaught; like any young animal it needs a supportive environment with caring, protective parents to enable it to grow and mature and live comfortably within its own abilities.

How could an open, explorative young mind not be damaged when it faces a high degree of hostility and non-acceptance, when crucially, at the same time, it is denied the defence of avoidance (freedom to run away)? What could the controlee's brain circuits do to accommodate this situation? How could the controlee's brain be wired to adapt to hostile penetrative interference? The answer, of course, is that it cannot, that is why it causes such damage, distress and mental harm.

Today, most societies endeavour to deal with the problem of mental ill health compassionately; however, as discussed in Chapter 10 there is the very troubling exception in the form of a universal imposition of compulsory education which, like the other recognized forms of abuse (sexual, emotional and physical), also creates long-lasting mental damage, even suicide. A child needs to have support, training and education to be able to deal with the heteronomous control that it faces throughout its lifetime, but to make this teaching compulsory (from a very young age) is a travesty of justice and a violation of the child's autonomy and rights.

All heteronomous controls are deleterious to the controlee, many are contradictory, unreasonable, painful, onerous and false, and they are enacted because the threatened punishment would be more severe than the cost of compliance, all of which means that heteronomous actions cannot be enforced and enacted without causing damage to the controlee's mind or body.

This results in distress, mental illness, self-harm, suicide, depression, anorexia, obesity, anxiety, agoraphobia, drug taking, alcoholism, addiction or antisocial behaviour, for all are ways to express, or to dampen down, the unrealistic or impossible heteronomous pressures or demands that have entered and overwhelmed the controlee's mind. The expression of mental distress, if only in a small (perhaps ultimately ineffective) way, restores an element of autonomy to the beleaguered controlee. For those who exhibit mental distress are in effect saying, (as they are overwhelmed by the lies, controls, deceits and directions of others) "If I conceal my true self within my own disintegration I will preserve the last vestiges of my autonomy, this is my last desperate ploy to govern a small fraction of myself."

When a child from a loving family, with adequate provision and care, is emotionally, physically or sexually abused she/he sets out on a mental journey that goes from having an untroubled existence to one (at best) of fear, anxiety and uncertainty. This situation is now better understood as a diabolical injustice that needs to be remedied, hence the perpetrators are now more likely to face justice than before, but that does little to help the victims deal with the fact that they have experienced a brutal loss of autonomy.

Now imagine this brutal loss of autonomy taking place at the beginning of hominin evolution: no protection from the police or army, no support from counsellors or friends, no legal redress and no verbal language to express your plight. That, broadly, would have been the situation that faced the early hominins and it is through their traumatized minds that evolution has selected a top-down mode of survival to mitigate the worst excesses of motor-control. What is now required is a deeper understanding (an unpicking) of the process of having to conform to heteronomous controls (of being heteronomized – a newly coined word) to see how the controlee must absorb each demand, respond to each demand and how every compliance involves a loss of autonomy.

At present, the controlee is forced into a position where she/he needs to integrate these necessary and burdensome actions of compliance with the considerable demands of her/his own body (noting that these inner demands of self's own body are all that the animal needs to attend to). We are all forced into developing a mental system (requiring extraordinary mental agility, deceit and an enlarged brain) that allows autonomous and heteronomous controls to merge within a single person. Integrating the inner needs of the individual with the heteronomous demands of the controller is at base an impossible task, it creates the imbalances and difficulties which determine whether the outcome for the controlee is mental ill health or normalcy (but it can never produce animal equanimity).

Heteronomous control has given rise to a situation where the controlee is no longer totally accepted as her/his self, ('as I am, warts and all'), as a natural autonomous being (in the way that animals accept each other as simply being there).

Once an individual is thought of as an object suitable for control, improvement, or violation then the controller will justify his/her behaviour by making out that the controlee is 'inadequate', 'unacceptable', 'a failure', or in some way 'undesirable'. All these pejorative terms are constructed by human controllers, so they can enhance their own position by manipulating the controlees. It is inevitable that the controlee then becomes subject to assistance and improvements directed by others (and by self) and this has created a situation where we have come to believe that interference can be justified (one of the many fallacies spawned by heteronomy).

Acceptance of self as an independent being (which is the opposite of being a subject for manipulation) is essential to a proper, healthy relationship between autonomous individuals. It is not simply that a person needs to be accepted as she/he is, that is, not in any need of fundamental change, improvement or rearrangement, but that it is also intolerable that unsolicited criticisms, wishes or directions from any source have the power to 'stick' in the mind of the controlee and generate change. Indeed, animals are accepted exactly as they are because no other animal can do anything to change them because their minds are sealed against top-down intrusions (and they do not have hands capable of manipulating the situation). Mental distress cannot gain a foothold in the animal mind for it lives in a world where you are accepted as you are and where you are safe from heteronomous control, so there is nothing to perturb autonomous, bottom-up functioning.

A mind that is organized on the pattern of the original animal model (always preserving autonomy) cannot remain normal, rational or pure (righteous) if it must develop a top-down ability to accommodate heteronomous controls to survive.

Controls have a tendency to escalate, to become harsher, compulsory and unavoidable, that is, they become intolerable, leaving the controlee's mind with no option other than to retreat into so-called 'abnormal behaviour' – to blow a fuse, become out of control, become non-responsive to others (depressed, manic, schizophrenic, suicidal, self-harming or anorexic), all of which is perfectly understandable and justified, because it preserves the last vestige of autonomy beneath the irrational facade.

Following the crisis, or breakdown, relationships are then determined (dictated) by the patient's (victim's) apparently illogical (but residually autonomous) behaviour, rather than the heteronomous power of the controller. Considering the difficulties (bullying, sexual exploitation, punishment or exam failure) that the controlee must have experienced or been threatened by before arriving at this point, this flight to irrationality should be seen more as a triumph of autonomous resistance rather than a descent into madness.

Once an accurate balance sheet is drawn up, showing a true record of the costs of heteronomous control, it will become overwhelmingly apparent (despite the alleged benefits) that the costs of heteronomy and motor-control are too great for the individual, society and nature to sustain.

The previous twelve chapters have mainly focused upon heteronomy. Next, we need to examine autonomy in more detail; it is the primary system of animal governance, yet it has been overridden (tragically diminished and dismissed) by the hominins because of motor-control. If we look towards the animals, we can observe autonomous governance in action and we can see that their intraspecific behaviour offers hope for us beleaguered humans. The mental stance of the animal is far more equitable than that of the human; it is the pattern that has been selected for every animal on the planet, except the hominins; surely there is something we can learn from this disparity?

CHAPTER THIRTEEN

AUTONOMOUS BEHAVIOUR OF ANIMALS

Wild animals live contentedly within their environmental niche and as they do so they deliver sermons on the perfection of autonomous existence which humans choose to belittle or ignore. Animals are reliant on food provided by nature; they are out in all seasons and in all weather conditions. They use no fossil fuel, wear no clothes, and follow no fashions. They use virtually no medication and no surgery; they have no schooling, attend no lectures and listen to no instructions. Animals do not undertake paid work, become slaves or slave masters, or prisoners or prison warders, they virtually never have sex outside of oestrus and they never sexually abuse the young of their own species.

Wild animals have no distractions (no phones, videos, television, books, art, drama, theatre, music, no gardens, no sport, no cars, no hobbies, no commemorations, no restaurants and no holidays) yet they remain calm and focused, not anxious, not in need of antidepressants, sleeping pills, alcohol, cigarettes, or recreational drugs.

How do they do it? How is it possible to function in this non-demanding way?

They are often short of food, their offspring have high

mortality rates and most species are subject to predation, yet wild animals for hundreds of millions of years have lived and survived long enough to breed and evolve. They have been able to do so because they have an enviable, equitable mental stance towards themselves, members of their own species and the planet. For example, everyday a horse has to find a large volume of grass to eat; this would be a worryingly arduous task for a human but the horse does not seem to be concerned, it does not plan for the future, it does not make hay or grow oats for the winter, it just eats what is available (provided by nature); if there is no food in the area (due to drought, flood or snow), it cannot eat, it simply stands quietly, losing weight until it dies. If grass does become available (say in the spring) then it eats and puts on weight and breeds. There is, always, an underlying mental stance that exhibits a confident reliance on nature and a belief that 'all will be well'.

The mind of the animal has a default position of 'emptiness' that waits for external and internal events to trigger responses selected by bottom-up programmes. Humans seek, always, to fill their mind with information. Our top-down minds are brim-full with details necessary for either becoming a trespassing controller, or a compliant controlee. Animals are not anxious or troubled about what to do next, or how it should be done, or how they are seen by others; their minds remain calm, equitable and confident.

This chapter compares this equitable mental stance (gifted by nature) of the animal to the human mind that is full of the agitated and destructive ways that humans behave towards themselves, each other and the planet. Thus, exposing the deep motivations that drive the hominin species for closer inspection.

The gulf between the autonomous animal and heteronomous human is so wide that it appears unbridgeable, that is why we have such difficulty understanding the 'stance' or the 'mode of the mind' of the animal. There have been some notable exceptions, a few wise individuals who, with their sensitivity to the animal, have laid the groundwork for a new way of seeing. Poets, writers, thinkers and artists such as Paulus Potter, Jonathan Swift, Walt Whitman, Rainier Maria Rilke, Jakob von Uexküll, Gilles Deleuze, Giorgio Agamben, Eric Santner and Joe Hutto, the man who lived (for a year) as a turkey, have pioneered a receptive awareness of what it 'means' to be an animal. This, categorically, is not the world of fluffy animals, toy dogs, Walt Disney, or brave dogs in the military being awarded the canine Victoria Cross. Nor does it involve issues of welfare, cruelty, putting collars on cats, training dogs to assist the blind, loss of habitat, vegetarianism or vivisection (even though they all need to be addressed). This is the gradual exposure to human awareness of the mental stance (equanimity, silent witness, or being there) that the animal adopts towards the world which, when examined, offers a glimpse into the mysterious separation of mankind and the animals.

We need to think about the animal in a completely non-anthropomorphic way (as Gilles Deleuze suggested). It is a project that Jakob von Uexküll encouraged saying, "This was an excursion into unknowable worlds." Equally it is an attempt to think about the human as an animal, free from the distortions of the anthropological machine and its promotion of human exceptionalism (as Giorgio Agamben and others are seeking).

If we try to understand more clearly the everyday stance that animals take towards the world we may start to see how they live without shame, responsibility, possessions, clothes, new ideas, words, speech, religions, governments, prisons, schools, or any other expression of heteronomous control. The anthropological machine promotes human superiority over the animals at every opportunity and blinds us to any understanding of this reality. Human exceptionalism has always set the boundaries for any discussion about animals and any suggestion that the animal experience may offer benefits to the human is usually met with derision or outright dismissal.

The human stance to the world is not the norm. We are an exception among over eight million other species that all function in a similar way to each other, based on autonomous bottom-up control. Because of motor-control, humans have been forced to function in a uniquely different way that enables them to deal with the vicissitudes of heteronomous control. Animals, without the means to easily grasp and hold, and without top-down mental control or speech, comport themselves differently. They live with virtually no ability to manipulate (physically or verbally) their habitat, their conspecifics or other animals, yet their lives are successful and contented. This 'other way' of being in the world is nature's norm and we should look to it for guidance and information because it is so well established.

The bottom-up mental comportment of the animal has been present throughout the whole of animal evolution and that must mean that it has considerable benefits and advantages. By comparison, the hominins, with their enlarged brain and top-down mental comportment, accounts for an exceedingly small fraction of evolutionary time. That is why we should look to the animal, for it is through them that we can start to see what the autonomous life looks like and how it manifests itself in daily life. These are lessons that will be difficult to learn, not just because of the blindness and stupidity brought on by human exceptionalism, but because autonomy always operates imperceptibly (silently) in the background, maintaining a very low profile.

When an animal is confined to a farm, a zoo, a research laboratory, or a domesticated animal is confined to a house, or a child is confined to school to undergo compulsory education, or a prisoner to a cell, or a slave to the workplace, then that individual loses a considerable degree of autonomous freedom. It does not die but its function is limited, it can eat if food is provided, it can defecate and urinate, but it is no longer a truly autonomous individual where its basic body programmes can be expressed whenever its safety/danger scales so direct. If the individual is then harassed sexually, or forced to comply with heteronomous instructions, its autonomy is further compromised. Humans are now in such thrall to heteronomous control that often they are only able to minimally express any truly autonomous behaviour or self-governance. As a result, our sensitivity to autonomy and freedom has become numbed to the extent that we are barely troubled by these examples, we simply do not understand the damage that is being done.

How has this shocking loss of autonomy been accommodated in our everyday behaviour? When autonomy is threatened or overrun, the individual's safety/danger scales will adjust to that fact, hoping (expecting) to regain full control later. Compromises will always be made when the situation dictates, for in dire circumstances survival is paramount. However, it should be noted that this is a relatively new situation within nature; autonomous function has virtually never been compromised (except at death) throughout millions of years of evolution up until the hominins violated this imperative and adopted their intrusive patterns of motor-control.

Autonomy is the ability to freely enact the programme that has priority at the top of the individual's safety/danger scales and, so it is the pivotal act of true self-governance. Wild animals live autonomous lives, whereas virtually all humans, virtually all the time, lead heteronomously controlled lives being conscious of, and adjusting to, the top-down demands, needs and wishes of others. Even basic bodily needs such as eating, sleeping, breastfeeding, urinating, defecating and sexual intercourse are riddled with problems of privacy, respectability, indoctrination, censure and taboo that originate from the restrictions and directions of heteronomous controls. Only reflex actions, such as breathing, sneezing and vomiting have remained truly autonomous acts (but once they are enacted they too become overlaid with heteronomous concerns).

The overwhelming background to the mental environment of humans is heteronomous control, which curtails autonomous expression. Heteronomous governance means that the controlee must enact the thoughts, wishes, desires of another person; this is a remarkable and dangerous ability that no animal (apart from the hominins) possesses.

Both autonomy and heteronomy have the power to change behaviour, but the individual can only remain autonomous if he/she is able to freely choose (bottom-up) how, and when, to respond to a given stimuli. Current explanations have a troubling lack of clarity at this point, for how can the individual be truly free if the behaviour is sourced heteronomously? Who judges the freedom, the controlee or controller? Are you free if you willingly attend a patriotic parade but would have been called a traitor if you didn't? Do you go to school 'freely' when attendance is compulsory? Does (can) the female really 'freely agree' to intercourse when she is in a situation where she is unable to establish anoestrus and the male considers intercourse as a conjugal right?

These questions expose a conceptual muddle, but two things are clear, first, most behavioural decisions made by the individual human are made because of heteronomous imperatives (which are not expressions of basic body autonomy). Second, it is not possible to fully comprehend the function of autonomy in the human without first understanding the ubiquitous presence of heteronomy. In the discussion so far, we have come some way towards a more realistic understanding of how heteronomous controls interact with autonomy and disrupt its original priorities. We are held in thrall to heteronomy and our personalities are constructed out of the heteronomous controls we experience, so not until we are able to set up an 'Archimedean point' that is genuinely free from heteronomy will we be able to understand the space within which natural bottom-up autonomy functions.

Immanuel Kant lacked the biological grounding necessary

to understand autonomy as a bottom-up concept; he thought autonomy was expressed (and defended) top-down by the conscious will and this is still the orthodoxy today. However, top-down responses arise out of heteronomous control, so they are all heavily contaminated with heteronomous inputs; indeed, they cannot be free of heteronomy. All of which is very far removed from the animal expression of a genuine bottom-up autonomy, hence it is rare to observe a truly autonomous reaction made by a human being. Mostly one sees the personality of the controlee (with her/his multi-faceted heteronomous content) trying to defend self by means of heteronomously learnt responses.

In the main, human governance stems from a mixture of heteronomous controls that (somewhat like the bottom-up system) have to have their own system of priority attached to them, so that those controls that come from the more powerful controllers, such as parents, headmasters, policemen, husbands, wives and the state, are attended to (complied with) first, which means that self's own more genuinely 'self-generated' activities tend to be the last to be attended to, or are maybe not attended to at all.

This necessity to be able to enact the top-down decisions of the will is thought, by today's philosophers, to be an autonomous act but it is not; it is simply a measure of the heteronomous control that the controller has over the controlee. It is at this point that humans have made (make) their most grievous error. Being free to choose which heteronomous control to enact at any one time should not be classed as a true expression of freedom or autonomy. It is simply part of a human defence to heteronomy, for once autonomy is lost we must be able to do what the controller (parent, teacher, state) wants us to do, which is not in any sense of the word autonomous behaviour.

Genuine autonomy is difficult to put into words (it predates words), it simply is just letting the body defend itself with its bottom-up programmes. That is all there is to it, hence it is either in operation (as it is with the animals) and there is no problem, or it is overridden (as it is with the hominins) and severe behavioural difficulties follow in its wake.

In the following paragraph I will mark, with a superscript numeral, the moment an autonomous priority decision is made...

At dusk the female badger wakes,[1] leaves the set,[2] grooms,[3] and moves off.[4] She slowly walks through the damp fields, nose to the ground,[5] she stops to root in the turf,[6] and finds a worm and eats it,[7] then she raises her head and listens out for other badgers.[8] The badger moves on to find more worms and then repeats the process,[9,10,11,12] then walks across to the latrine site near the hedge and defecates.[13] She then moves back into the field to find more worms and insects,[14] then back to the hedge again to find berries,[15] then she smells/hears[16] the farmer entering the field and, sensing danger,[17] runs off,[18] keeping close[19] to the hedge that provides cover, not taking any risks by stopping to look for food.

Here we see the badger autonomously acting out patterns of behaviour that have been moulded by selection over millions of years of evolution, but it is in the changes from one unit of behaviour to another that we see autonomy in action. Only at this moment of the change do we see the switching action that is the essence of autonomy. The moment programmes are switched ([1-19]) we witness the individual exhibiting genuinely autonomous control of its behaviour.

The individual badger is a living package that is prepared for all badger eventualities. The badger can switch its behavioural programmes autonomously so that it sets in motion the bottom-up programmes, provided by evolution (nature), that ensure that the most pressing dangers are always met first. The selection of one programme rather than any other is the point where the autonomous control of the individual is exposed to view. When the switching mechanism has several easily achievable options (of similar priority) at its disposal (say, move up the field, or go into the hedge, or move to next field) then governance is relaxed, the individual is in a 'comfortable' position. As danger increase, so the threat to autonomy increases and eventually the priority options diminish down to two – flight or fight – and the governance necessarily becomes more focused on immediate survival.

The badger mind is not tasked with having to work out what to do in every one of these situations; it simply stores a repertoire of activity and then facilitates its switching mechanism to set in motion the discrete unit of behaviour, that has been available to it from birth, which then becomes the next pattern of activity. Thus, the badger mind is free of any responsibility for generating the response top-down (unlike the heteronomous mind of the human) and so the conscious self of the badger can exist, float, glide or relax in the certain knowledge that it has nothing to 'do' (nothing to fathom out, nothing to create or think about). It simply inhabits (exist within) a calm space in the mind, which is the essence of 'being'.

Like all animals the badger does nothing other than that which comes into its brain bottom-up, it is free of responsibility and it has no duty, it is not tasked to achieve anything for it is solely the free flow of the autonomous switching activity of the safety/danger scales that sets in motion the next unit of behaviour.

The existence of animal autonomy is difficult to discern; it functions silently in the background, switching from one bottom-up programme to the next in a seamless continuum. If it is not apparent, then one can guarantee that autonomy is functioning freely. When autonomy is unable to function efficiently then its existence can be inferred from the individual's behavioural agitation. When you see difficulty, anxiety or unease you see an autonomy that is having difficulty implementing its priorities (or, with humans, you see an autonomy that has already succumbed to heteronomy).

When autonomy faces conditions of persistent heteronomous

control it cedes control to the controller and the controlee is forced to act in ways that are very different to when it acts freely. Autonomy under heteronomous control is no longer a true autonomy (an autonomous autonomy); it is a divided, compromised autonomy.

Persistent heteronomy creates a new system of governance in the controlee's mind that is quite different from the autonomy that governs an animal's behaviour. Animal autonomy is strong, powerful, capable and confident it can always be relied upon to deliver a satisfactory bottom-up response, which never necessitates any outside assistance. (The autonomous bottom-up defences of the animal only fail in extreme situations, such as predation, ill-health or injury, and on some other rare occasions such as floods, tsunami, volcanoes, quicksand, tar pits or briar patches.)

Wild animals are not obedient to others (they do not know how to be obedient), they may briefly exhibit submission, but never obedience. Domestic animals are trained to become obedient. Some are taken by their owners to obedience classes where, the more obedient the dog, the more prizes are won and the prouder the owner becomes (all without a hint of shame). Winning a cup for obedience is a horrific insult to the dog's true nature but humans, weighed down by heteronomy, are blinded to such understanding. However, there may be a faint glimmer of hope in all this, for it is possible that in putting dogs on leads and training them to be ultra-obedient, we are subconsciously seeking to obtain an insight into the obscenity of our own tightly coerced and controlled lives.

Some domesticated animals, for example, working horses under harness, must to some small extent operate top-down – the ploughman shouts "Go," slaps the rein and the horse moves off. The shepherd blows a whistle and the sheepdog moves to the fetch command. Dogs for the blind have been trained to behave abnormally and not to become interested when another dog or bitch walks past. All in their different ways are subjects of heteronomous controls that have been inserted into their minds by humans (not by their own species) in the training process – they are all actions that certainly do not originate in autonomous bottom-up behaviour patterns of the wild horse or dog.

Interestingly this is in some respects a similar process to that experienced in the early stages of hominin development. Once domesticated and under control of another, the controlee responds to commands (complies) because she/he has no other option. When we witness animal domestication and training, we are witnessing something that is similar to the protohominins falling victim to motor-control, losing their autonomy and becoming subject to heteronomous controls, except that we, the hominins did (do) it to ourselves.

13:1 How the animal mind differs from the human mind by being impervious to heteronomous inputs:

Psychoanalyst Dr. Peter Hobson of the Tavistock Clinic, London has commented, "that even though he has learnt to analyse the countertransference (which enables him to understand how the human client is making him feel) he is unable to 'understand' or 'connect' with animals. The feral animal does not present (allow) any part of its mind to be engaged (contacted, connected) by humans". Or, as Walter Benjamin said, "There is no addressee of communication."

My own experience with animals is similar. When I groom the horse, walk the dog or stroke the cat I try to communicate with them, I have a 'bond' with them and they have a 'bond' with me (probably based on the fact that I feed and shelter them), I can train them to the harness or to come to heel (can't train the cat) but, over fifty years, however hard I have tried, I have not been able to get inside an animal's mind so that it responds to my enquiries, my entreaties. Compared to humans, it is as if they are 'not at home', but I now see this as an example of the protective benevolence of evolution, for this 'indifference' on the part of the animal is the mechanism by which the autonomous mind is protected from the dangerous penetrations of unsolicited heteronomous control.

Animals do not enter a top-down mental dialogue with others; there is no need to have cells in their brain to enable them to respond in this way and if they had, it would be a dangerous weakness. Thus, the animal brain remains empty of heteronomous inputs, whereas the human brain is full to overflowing. The human mind keeps searching for the cause of its underlying discontent, it has a relentless, restless longing for better, safer circumstances; in contrast, the animal mind is content and at rest. Put another way, animals have not evolved a neural route for direct heteronomous mental contact because they have no need to make (there is no benefit in making) contact in this way. In effect they are saying there is no sufficient reason to engage with the mind of another animal given that it possesses an inherent threat to my autonomy. A self-sufficient autonomy needs no top-down contact; how could one animal assist the autonomous function of another animal? Only those who are subject to motor-control and heteronomy have the need for such penetrative mental contact.

Only the heteronomously obsessed human would be so arrogant, so lacking in respect or empathy, as to intrude upon the behaviour of another independent individual. (Yes, lacking in respect, for how can one truly understand the complex requirements of another individual at any given moment in time?) Bottom-up behaviour is selected autonomously according to basic inherited programmes. With heteronomy the top-down mind creates and assembles the behaviour it enacts; nothing is set, fixed, or laid down, everything is fluid, every action and event can be examined and changed, nothing is certain, everyone has a need to continuously improve their defences.

I may wish to try to converse with the family pony, reassure her, scold her, but why would she (or the cat or the dog) want to speak to me? There is no selection pressure on a contented autonomous pony that makes it want to get into mental contact with me. It makes no sense for animals to seek mental contact with humans; it would be like entering the world of a schizophrenic, an invitation into a lost coherence. For, by what aberration did I scold the pony to correct her behaviour when I was the one that had confined her?

For animals to understand and converse with humans they would first have to learn a 'foreign language' that would enable them to exist when in thrall to another. They would somehow have to be made to experience, and then fear, the world of motor-control. Not simply the wearing of a harness or the use of the lead but also the experience of sexual assault when anoestrus. Some domestic animals are halfway down this path: cows endure the vet thrusting an arm into their rectum, or inserting insemination straws into their vaginas, or vaccination needles into their muscles, or vermicide down their throat, all penetrative actions that wild animals simply would not tolerate (hence they need to be sedated if they are captured for 'treatment').

When examining autonomy, it is necessary to ask what acts (other than 'being there') could a conspecific enact that would truly assist another's autonomous existence (long-term self-governing survival)?

Where do these beneficial acts begin and end, why would it not encourage a dangerous dependence or a dangerous interference (as it has in humans)? We need to understand that it is solely the experience of heteronomous controls that give rise to this 'need for assistance', for the victim of heteronomy is unable to cope on her/his own whereas the autonomous animal is happily self-sufficient.

Autonomy is autonomy, it does not (cannot) want assistance, if it receives assistance then it axiomatically loses autonomy, or a degree of autonomy. (We know so little about autonomy, can it come in degrees?) Autonomy is individual governance; it has evolved to become the guiding principle of mammalian life; all behavioural programmes have evolved to maintain autonomous existence. When protohominin autonomy entered into a Faustian pact with motor-control it suffered a tragic emasculation by falling under heteronomous governance. Perhaps humans should see their loss of autonomy in a similar way to those who tragically have lost their sight or a limb – they have survived but have paid a heavy price. So too has humankind, for as a species we have paid a colossal price; we have lost the ability to function autonomously.

An autonomous animal is mentally self-sufficient, it is single-minded, (is a singularity) it survives under its own governance without any assistance. The animal mind basically does not engage with the minds of others, it is indifferent, it is not accessible, why should it be? It is autonomous, it can govern, control and organize itself by itself, it does not need help from friends, vets, politicians, priests, it is already programmed to feed itself, avoid predators and other dangers and to breed. If autonomy appears to seek help or assistance then it is not acting autonomously, it is under the influence of heteronomy.

Heteronomy speaks verbally, autonomous animals do not 'speak' verbally, animals convey their true body state and emotions by such means as body posture (alert, relaxed, tense), with odours that can readily be read by others in the vicinity and by various other patterns of behaviour. These body signs reflect the true situation and in the main they do not lie. Top-down heteronomous controls lie from the moment of inception. No individual wants, desires or needs heteronomous controls but is coerced into tolerating them even though they are a violation of truth (an injustice) from the moment they override autonomy. As the primatologist Dian Fossey said, "I feel more comfortable with gorillas than people. I can anticipate what a gorilla is going to do, and they are purely motivated."

13:2 Thoughts about animals:

On Epsom Downs, England, a few years ago, a man became the owner of a Derby winner. He had purchased the horse, as part of a business deal, just a few months before the race. The night of the victory he threw a lavish party to celebrate and for the next few years he enjoyed the boost to his status that is bestowed by being the 'owner of a Derby winner'.

He was able to bask in this glory, yet he had not run the race, nor had he bred or trained the horse, he had simply purchased it and owned it on the day. Meanwhile, the horse, unaffected by his exertions, did not display the trophy in his loose box, did not want a slice of the winnings, did not have a party with his stablemates that evening and next morning did not pull rank as 'the Derby winner' when he passed a shire horse in an adjoining field. Nor did a horse in the next village say, "My brother has won the Derby."

Winning meant nothing to the horse and his response was to continue to eat hay, drink water and defecate. Like all other animals he did not crave ownership, status, celebrity, victory, prizes or stud fees. Such things are of no interest to the animal, they only make sense to the human because they are part of the essential defensive facades (personalities) that we each erect as we struggle to survive in a world of motor-control, coercion and heteronomy.

This illustration fails to encompass the full picture for this was a domesticated horse, it has been bred to accept fields, fences, stables, bridles, saddles and leading reins. A wild horse would not have tolerated any of these things, and would not even have set foot inside the racetrack. So, the true comparison between mankind and animals is more extreme, but the example starts to illustrate the adjustments that humans will have to accommodate if we are to understand the relationship that animals have with each other, the crassness of the relationships that humans now have with each other, and by comparison to see the relationship that humans 'should' have with each other.

It is the imposition of heteronomous controls that creates the major difference between the organization of the animal mind and the organization of the human mind. By examining some of these differences we can start to see just how much of our daily mental activity is consumed by the directions of, and responses to, motor-control. To expose this gulf, I will list some of the things that animals do *not* do (because a fully autonomous animal has no need for such things). Describing autonomous behaviour in terms of what it is not is somewhat indirect but if the reader reflects on what animals do *not* do they may, by implication, see that we should not be doing them either, thereby glimpsing the extremely problematical nature of our own behaviour.

13:3 Listing some of the differences between animals and mankind, highlighting the gulf that motor-control and coercion have created:

1. No animal puts another animal on a lead, or in a harness, or ties them to a tree with ropes. Animals are never caught and held against their will unless they fall into the jaws of a predator, sink into a bog, or become trapped in a thicket whereas hominins can be grasped and held virtually at any moment in their life. Animals do not domesticate other animals, break them in, ride them, race them, or use them as beasts of burden or demean them by turning them into pets. No animal trains another animal to be obedient to the controller. No animals perpetrate genocide or engage in full-scale war. Animals do not behead, hang, stone, electrocute, shoot, rape, torture, starve, entomb, imprison or detain their conspecifics. No animal is subject to, or responds to, heteronomous control. No animal

can be conscripted. No animal can gain (top-down) entry (or any unsolicited entry) into another animal's mind, whereas humans are in and out of each other's minds every minute of their day.

2. No animal is overly concerned with its own weaknesses, such as, lameness, wounds, coughs, mange, parasites, wall-eye, odours, or dung on the tail (or nudity), all are accepted passively, as are age, sickness, or starvation, as well as the need to defecate, urinate and pass wind. Neither are they concerned about the fate of others, cattle will graze contentedly outside a slaughterhouse, as W. H. Auden wrote, "dogs and horses get on with their lives as terrible events unfold." No animal seeks perfection, or beauty, or considers aesthetics. No animal wears clothes or marks its body with tattoos nor do they brand other animals. Animals do not measure their breasts, girth, height or weight. No animal cuts its own hair, wears wigs, has hair transplants, plucks its eyebrows, or shaves its face, armpits or pudenda (even when it has mange or lice). No animal applies make-up or body paint or has fashions. No animal performs castrations, circumcisions, or clitorectomies or any other surgical procedures upon themselves or any other animal. Dogs and rats do not, for example, use chimpanzees for experimental purposes, nor do they put cats in cages to administer electric shocks (nor do animals protest about the experiments that are carried out upon their relatives, that is how we get away with it). Animals do not put other animals in zoos. Animals do not attack and damage the females of their species. No animal species prostitutes its females. No animal carries a gun, knife or cosh. No animal uses bombs, drones or guided missiles. No animal drives a car, flies a plane or builds roads.

3. No animal species has royalty, emperors, presidents, popes, generals, admirals, mayors, and orders of chivalry, freemasons, doctors, veterinary surgeons, lawyers, accountants, bankers, teachers, government officials, politicians, judges, policemen, prison officers, entertainers, sportsmen, artists or actors. No animal presents medals, cups, certificates or bestows titles. Animals are not proud of each other (they never say "I am so proud of my children, my staff or my country"). Animals do defend their local territories, but they do not have flags or nation states. They do not stand to attention, or salute, to show respect for their parents, teachers or officers. No animal shows allegiance to their country. They do not have badges, uniforms or rank. No animal is trained to express honour. No animal is obedient to the orders or directions of another. Wild animals are ungovernable – that is the essence of their beauty; they govern themselves. There is no need for top-down democracy in the world of bottom-up, autonomous individuals, as all animal societies demonstrate.

4. No animal species has a system of compulsory education. No animal has formal education, schools, colleges, universities, examinations or academic failures. No animal takes or gives instructions. No animal sets out to improve performance. No animal tests another's knowledge or sets examinations. Animals are born with evolutionary knowledge, wisdom and an ability to learn and update memories, which are expressed through their bottom-up programmes. They have no need for overt (top-down) teaching, they learn from experience (sight, smell, taste, touch and hearing). No animal has any qualifications. No animal suffers the demand, "Do this," "Learn this," "That's wrong," "You have failed," "Come here to be punished (or humiliated)." Or the reverse, no animal is praised as successful, brilliant, top grade, selected or elite, they

do not say, "You are a credit to your parents." No animal asks questions. No animal beats the spirit out of its offspring with cane, switch or tawse. All animals evolved prior to the word; they do not need speech, writing or books to communicate, which makes it much more difficult for them to lie (they can be deceitful on rare occasions whereas humans deceive each other every day). No animal has words, alphabet or numbers, yet they can count and communicate without any need to make unsolicited penetrations into each other's mind (or perhaps this should be the other way around – they communicate very efficiently *because* their minds are not full of the dross of heteronomy.)

No animal names or numbers their offspring, they know each other by look, smell and sound. (A name is a shortcut used by humans for the efficient heteronomous manipulation and social control of others.) Animals do not act, paint, draw, sculpt, or display artefacts in galleries. Animals do not write poems, plays, novels, they do not have a culture. ("Culture is a mad flight to deny that we are animal," said an artist talking on television in 2008, whose name, unfortunately, I did not record.)

5. No animal has any religious beliefs. No animal worships any god. No animal celebrates any religious festival; they have no religious texts, scriptures or buildings. No animal goes to church, temple or mosque. No animal has marriage rites or funeral rites. Even those that pair-bond do not have ceremonies with invited guests. No animal wears a wedding ring or swears fidelity and obedience. No animal records and celebrates birthdays. Animals do not celebrate anything, nor do they have remembrances. Animals can forget things because others will not hold them responsible. Animals do not laugh, or cry and they have no attachments (apart from family).

6. Animals do not seek 'the good life', they are content with the one they have. They do not have holidays to relax and unwind and generally they are not waited upon by others. Animals have no shops and they do not have commodities. They do not need to be entertained; they have no theatres, sports stadia or television programmes. They do not seek luxury. No animal has money, nor do they have slaves or sex workers, nor do they work for the benefit of another. No animal needs to work to be able to buy the food essential for its survival. Animals do not sell or buy anything (certainly not their bodies). No animal goes shopping. No animal cooks meals or goes to restaurants. No animal is an employee, no animal is an employer. No animal has possessions (they do claim temporary 'ownership' of nests and territory). Animals are neither rich nor poor. Animals do not pay rent, write cheques or pay taxes. Animals do not accept, or deliver, praise or blame.

   Animals are not minded to act in any of the ways in the above list; they are indifferent to the attraction of these activities because they are spared the need to build defences to motor-control. However, they live perfectly satisfactorily lives whereas humans struggle to cope with the rise and rise of heteronomy. Intraspecific violence, rape, torture, slavery, concentration camps, psychopaths, criminals, ruthless dictatorships are not examples of the animal nature of humans – these examples of intraspecific violence are quintessential examples of the 'human' nature of humans.

13:4 Animal acceptance of the given:

Animals exhibit an acceptance of their individuality, as it is, in its present state, with no qualification or conditions, indifferent to any need for improvement, exhibiting an inviolability as they live out their life in the present. This stance underlines the security that comes from the ability to defend autonomy and it gives rise to their mental state of equanimity (which, I suggest, is the same as the experience of eternity, bliss or enlightenment that today is earnestly sought by all the major religions of the world). It is gifted, by nature, to each animal and prominently displayed by them, yet it is unseen by humans and unavailable to them. In answer to the question posed at the beginning of this chapter ("How does the animal do it, how do they function in such a non-demanding, contented way?"), we can now see that when animals are at rest their minds return to a state of tranquillity and equanimity. This is quite different to the human mind, which is continuously disturbed by the agitation caused by heteronomous control.

The default setting of the animal mind is peaceful (blissful); it reflects a primordial tranquillity that arises out of non-activity. It is probable that while the bottom-up programmes are functioning smoothly this default also becomes a standard background set. It is even possible that this remains so for all the animal's life, for apart from serious intraspecific confrontations or attacks by the predator, there is little else in an animal existence to disturb its equanimity.

Animals have been crafted out of evolutionary forces that select those most fitted to survive in their environmental niche. The frog in the pond, the weasel in the hedge, the adder on the heath, the warbler in the marsh, all live out their species lives at ease within the open. Animals are born with their species abilities, so they are not tasked to (anxiously) provide them for themselves, which means they have no need to be concerned about any top-down issues. The essence of the animal (although they are unaware of the fact) is absolute trust in evolution (similar but more powerful than those who have trust in God). The animal is indifferent to anything that is not programmed bottom-up; they simply exist and verify their existence in their being.

For animals, the process of living and dying are solely bottom-up activities. The conscious mind of the animal knows no other than to accept and respond to the bottom-up programme that is currently selected. There is no checking or worrying to see if the response is 'correct'. If it has been selected from within the bottom-up programmes it must be correct. This is the case even with the act of dying – dying is always a bodily truth.

Death is just another (or the perfect) example of bottom-up action (or lack of action). In the animal there is no top-down activity that is tasked to do anything about death or dying (there are appropriate bottom-up moves such as lie down, rest, seek shade or hide from predators). Bottom-up responses are not tasked with any anxious or futile manoeuvres (such as praying or seeking a veterinary surgeon), the animal is as accepting of the givenness of death (when it comes) as it is with life. This is how the animal deals with its own impending loss of life, be it the victim of predation, the runt of the litter, the insurance chick, the cubs that are killed when a new male displaces the old alpha male, or death through age, injury or starvation.

All animals accept the fact that if their bottom-up programmes do not have an adequate response to a situation then they will expire (and they do so in large numbers), their conscious mind is not tasked to alleviate this outcome. Here is where the human top-down organization of medical science needs to be viewed from a wider perspective, for given that we are all destined to die, is the ability to extend life (for a limited time) worth the suffering inherent in a lifetime of heteronomous controls? Presumably this is what is meant by the phrase, "The wages of sin are death," (Romans 6:23); animals do not 'sin' (do not experience the trespass of motor-control) and so they live as if they think they have eternal life, that is, they have a mental stance that accommodates death as readily as it accommodates a cold day – the animal mind is never responsible for any outcome.

Bottom-up animal behaviour is predictable, because it has been enacted and observed in the same environmental niche for millions of years. Hence, animals 'understand' and are familiar with each other's 'species tramlines', which makes their behaviour easy to read. In *Being and Time*, Martin Heidegger describes animal (bottom-up) behaviour as, "being poor in the world," but paradoxically it is also the reverse. It is true that animal behaviour is limited – fish cannot graze on mountain slopes, rabbits cannot climb trees, pigs cannot fly. Humans may class this as restricting but evolution operates within discrete environmental niches – why would a fish benefit by being able to graze on a mountain? This so-called 'poverty in the world' is simply the outcome of the laws of evolution and the paradox is that the animal that is said to be 'poor in the world' is the one that exhibits contentment and equanimity, whereas the allegedly intelligent, free and 'rich in the world' human, at the pinnacle of evolution, has lost all genuine contentment and equanimity. We all have a persistent anxiety, due to our fear of motor-control, but we do not understand that fear, which means we are left with a constant need to find or construct something (anything) onto which it can be projected. In discharging this tension, we also explore our own angst and that will continue until we are able to find a way to restore autonomy.

13:5 The unitary mind of the animal and the divided mind of the human:

All animals have a unitary self; they are a singularity, a self-sufficient unit that rarely seeks (or provides) assistance. No animal can lose control of its own mind (apart from injury or disease) or have its behaviour directed by another from the outside. No animal can take over another animal's mind because it has no way to restrain that animal and enter its brain. The animal is always able to initiate avoidance and so its mind remains self-contained to the extent that heteronomous control is not even a potential threat.

The poet Rainer Maria Rilke, in his *Duino Elegies*, suggested that, "animals enjoy direct access to a realm of being – 'the open' – which is concealed from humans by the workings of consciousness and self-consciousness." This access to 'the open' is the mark of creaturely life, it is the standard experience of all animals irrespective of the species tramline they are programmed to run along, but it was denied to the human following our loss of autonomy and as a result we developed our top-down self-consciousness – that characteristic, conspicuous, self-revelatory sign of heteronomous control.

Animals look out onto the world in a straightforward (non-manipulative) observational way with a total acceptance of the environment and of their own and their conspecifics' behaviour.

The animal mind operates at the pace of the bottom-up programmes, whereas the human mind has top-down capacity to deal with (and recall) virtually any heteronomous control or instruction that the controller may choose. Animals are programmed to attend to their safety/danger scales; to eat, sleep, defecate, breed, and maintain safety, these species-specific programmes are linked together in a flexible way so as to be able to concede priority to the most pressing need.

In the moments between pulses of activity there is nothing to occupy the animal mind, absolutely nothing (apart from being subconsciously alert to the sound, smell and sight of danger). This is the default position in which equanimity, 'hereness' and contentment come to the fore to bathe the animal mind (this is not only for the odd moment, or an hour, for it can last for a long period such as the whole of a winter night), for if there is nothing to do, then the bottom-up programmes are at rest.

These spaces between events remain empty for the animal

but humans fill them with thoughts, questions, speculation, fantasies, fears, anxieties, loves, or with plans for self and plans for others, all to negotiate the rising tide of heteronomous control. In a normal day at work humans are bombarded with a constant flow of information, they are faced with having to make difficult decisions, under near-impossible time constraints, cooperating with people, some of whom they actively dislike.

Contrast this with a time when humans (perhaps) experienced a mental stance that in some inexact way comes nearest to that of the animal. Imagine a perfect holiday paid for by others, where you have no responsibilities, your phone has been switched off, there are no television, radio or computer links, all food and facilities are provided. In this situation you walk around, look at the view, swim in the sea, lie on the sand, eat a light lunch, and take a siesta. You observe others but there is no pressure to socialize, certainly no pressure to make plans, or work. You are reacting to your bottom-up body requirements, swim when hot, eat when hungry or sleep when tired.

No doubt human worries and anxieties would still be present in the back of your mind but imagine (unassisted by alcohol, drugs, music or videos) what it would feel like to be without all the normal everyday human tensions, and this, I suggest, is the nearest that we can imagine to the animal stance to the world. You may object and say that a holiday is an artificial situation, which is true, but observe an otter swimming in the river – it seems to me that nature provides the otter with all its needs over its entire lifetime, which in total amounts to much more than that which is being gifted in this example of a short holiday. Hence the unitary mind of the otter can continuously bask in a state of equanimity, living (resting) in the present moment, confident that its needs will be provided by nature. It is ignorant of the wider context, yet it has been provided with all the necessary bottom-up programmes and all the necessary external inputs to meet its needs because a species can only evolve into, and survive within, a niche that is bountiful and stable enough to supply these needs over long, tried and tested periods of time.

The human is always alert, watching for the dangerous controls and restrictions that others are eager to impose on them, and at the same time, ready and waiting to impose their own controls over others. Yet on odd occasions it is possible to experience a fleeting state of mind that allows one to see and understand everyone and everything in the world from the perspective of 'unconditional acceptance' (love?), where there is no criticism of anything and no need to improve anything (certainly no need for top-down defences) – this is total acceptance of everything 'as it is'. This glimpse is probably the nearest that humans come to an understanding of the animal, of what it feels like to bask in a state of 'being' or 'hereness'.

13:6 The animal experience of 'hereness' and equanimity:

The animal mental stance or outlook to the world appears to be the same for all extant (and presumably extinct) species, from hippopotami to newts, from seagulls to hedgehogs, from gorillas to weasels.

This outlook can be described as having confidence in their own presence, they are not consciously responsible for finding their next response, yet subconsciously they are not in any doubt that they are capable of being able to enact a suitable response to any situation, at any time, because it will be provided by their bottom-up programmes.

Animals remain within themselves at all times, they are serenely immutable, confident of their own behaviour, they do not seek alternatives, they are impervious to heteronomy, and they are content to exist within a state that (to use words used by humans seeking enlightenment) may be called 'hereness', 'taking place', 'such as it is', 'being' 'being thus', 'being such as it is', 'what one is', 'the calm state of wu wei (no action)', 'resting in the grace of the world', 'free from desire', 'taking no thought for the morrow', 'in the moment right now', Spinoza probably coined the best description, 'being at rest in oneself'. If one is at rest within oneself then the outside world, whatever its state (extreme or temperate), will be experienced as paradise for it is unable to disturb the equanimity of the mind.

All animals (except the hominins) share this stance to the world – relaxed and dwelling in the present moment – and it determines the individual's underlying feeling of equanimity and non-responsibility, which is reflected in its demeanour towards itself and its environment. This stance is beyond time and history, in the sense that the animal has no interest in history, nor is it anxious about the passage of time. This is where Jean Luc Nancy's 'source of sense beyond the human world' can be found, that is, it is not beyond this world (in the sense that it is up in heaven, beyond nature) but it is beyond the heteronomous human world (beyond good and evil), which means that this elusive 'source of sense' is closer to hand than we think for it is constantly present in the mind of every animal.

All animals exhibit a total confidence in their own abilities and they have no difficulty maintaining a positive life force. They do not seem, in any way, to doubt their own claim to existence – they are confident in the fact that their given deportment, be they small, large, bold, timid, fast or slow, is fitted for their survival. Animals, basking in their state of equanimity, appear to confirm this, for they seem to live their lives, within their environmental niche, with their minds confidently poised at the centre of their universe. There is very little reference in the literature to this animal stance to itself and to nature. We need to examine it more fully and in so doing gain an insight into the strangely distorted views we have of ourselves and the world.

Loss of autonomy has left humans with an inner uncertainty about their own presence. We have lost our original animal 'presence of hereness' because we are inherently unsafe, and we must seek answers here and there, past, present and future. This has resulted in us exhibiting an 'everywhere-ness', or an 'all over the place-ness', or a 'can be made to do anything-ness', which leads us to be much more preoccupied with past and future events than are animals. Animals do not record, study, commemorate past events or conjecture about future events; other than in an environment of top-down heteronomous control, it would be remarkably inefficient to do so.

Animals do not have the top-down neural circuits with which to manipulate the information in their mind, hence they live their life in the present moment responding, reacting or ignoring events according to inbuilt species programmes.

Humans need to understand this ability of the animal to live in the present for within it lays a tranquillity and mental ease which we rarely experience. If we were safe from conspecifics (in the way animals are) we would be able to let the world 'come to us' for there would be nothing to fear (no defences to plan); our bottom-up programmes would be ready and waiting to deal with virtually all of the species-specific dangers that the individual has to face. Humans cannot allow the world to 'come to them' on its own terms for they have inadequate inbuilt defences to deal with motor-control, so they must search top-down for better defences and construct cooperative alliances, and this is mainly carried out by the manipulating words, exploring the past, thinking about the future, looking for clues and testing any conceivable improvements.

Animal 'hereness' results from just contentedly doing what the bottom-up programmes dictate at that moment of time; this may have elements dealing with the past (say, memories that food is to be found in the next valley), or events in the future (say, migration), but it is minimal, there is not the desperate preoccupation with past and future that humans now exhibit as they mark family anniversaries, historical events and financial and employment routines. Hereness arises from acceptance of the current bottom-up behaviour and not wavering from enacting its requirements here and now.

Consider the stance that the animal mind has towards the world. A cat lying on a wall in the sun, or on the prowl, or motionless before pouncing on a mouse, or discreetly defecating or grooming – its mind remains calm, not agitated in any way, measured, confidently in control of the situation. No animal has thoughts of improving its own safety beyond that laid down in their bottom-up programmes.

In developing a top-down mental alertness to the dangers of heteronomy, *Homo sapiens* have paid an incalculably high price, for they have lost their ability to experience the equanimous state that bathes the minds of all animals. For how can the hominin mind remain in a state of equanimity when it can be penetrated, overridden and controlled as and when a controller so chooses? In being held, coerced and heteronomously controlled, we have become puppets in thrall to others.

We fail to see that the untrammelled animal has an effortless grace and equanimity that is lacking in the human. We fail to see that evolution bestowed the incalculable gift of autonomy on all the animals yet withheld it from mankind. We fail to see that it is an integral property of heteronomy that (initially) the controlee (victim) always 'fails to see' – that is how the walls of our entrapment are constructed. They will need to become apparent /transparent before we are able to be reconciled with our animal nature.

What mankind cannot bring itself to say (cannot bring itself to the point of 'enlightenment' to say) is that it is the animal that is divine and the human that is profane and corrupt. Animals do not realise that they live autonomously, humans do not realise that they live in a world of heteronomy, neither have experience of the other's situation. For this to be resolved, mankind will have to find a way to return to its animal nature for it is inconceivable that the autonomous animals would voluntarily succumb to heteronomous controls.

An animal can depend upon the fact that it has inherited a complete set of species-specific bottom-up programmes, which are subconsciously and autonomically modified and updated throughout its life. These programmes are alert to all eventualities and they can trigger actions that deliver the correct response to any circumstance that the individual is likely to encounter within its species niche.

Hence all behavioural decisions are non-problematical because they rise effortlessly out of the bottom-up programmes. This means that the conscious self of the animal has no responsibility, tension, anxiety, effort, or executive involvement in constructing those responses. All of which allows it to relax and observe, for it 'basks in', or 'floats upon', the enactment of the chosen priority behaviour of its bottom-up programmes. In terms of worry, work or organization there is nothing whatsoever for the conscious mind of the animal to 'do'. Nevertheless, all will be well because the species-specific, bottom-up behavioural programmes can be totally relied upon. It is this confidence, bestowed on them (gifted) by evolution, that gives animals their equanimity which can be seen in their poise, composure, calmness and inner control.

Acceptance of responsibility and loss of equanimity are discussed in the next chapter.

Endnotes: Chapter Thirteen

Endnote 1

This chapter suggests a repositioning of the human species within the natural world with the desirability of a full return to nature and autonomy. In so doing it ignores the fact of today's high population levels (overpopulation) which render any return to nature impossible (at present). The nutritional requirements of seven billion people cannot be met naturally (it is possible that humans have already reached unsustainable population levels even with agricultural production). Animal population levels are balanced to the available food supply by means of starvation and/or failure to breed. For there to be any chance of humans adopting a more natural existence, population levels would have to fall considerably, perhaps to the levels seen 5,000 years ago; only then could we return to living and feeding within the gift of nature. None of which precludes the fact that we could, should, need to, start to eliminate all incidents of unsolicited motor-control from the human repertoire.

CHAPTER FOURTEEN

HOW WE CAME TO ACCEPT RESPONSIBILITY FOR OUR OWN ACTIONS AND AS A RESULT LOST OUR EQUANIMITY

We have some inner understanding of how the top-down brain functions because that is the way our own minds operate but the animal experience of having a brain that functions bottom-up is less easy for us to understand (humans do have a basic bottom-up core to their mind but to a large extent it is overridden by the top-down system). Nevertheless, we need to examine the bottom-up experience more closely since it contains within it signs, clues, information and modes of operation that, by contrast, help to illuminate the human condition.

14:1 Puppets in the land of heteronomy:

The hominin brain has evolved to such an extent that humans, with their agile flexible minds, are now capable of remarkable feats of overt memory recall, planning, abstraction and articulation that no animal can match. In this chapter, I seek to contrast these superior mental abilities with those of the animal and show how, in coming to terms with motor-control, a Faustian bargain was struck in which the price paid by the hominins for these prodigious abilities was the loss of their greatest animal asset – their equanimity.

We are largely unaware of the consequences of this bargain because we have never experienced the calmer bottom-up brain state that was sacrificed in the development of top-down mental organization. There are hints of its existence in the scriptures and the work of some poets but apart from that little else. However, animals have been spared these top-down changes so, by careful observation, it should be possible to discern an outline of what it is that they have retained and by comparison come to see what it is that we have lost.

Counter-intuitively, this means we are looking to understand how the animal comports itself in a way that is superior to the human. I hear cries of incomprehension coming from the human exceptionalists, but it is only by comparing the stance that the animal has towards its behaviour with the stance that the human has towards his/her behaviour that the full implications of this tragedy will emerge.

All unsolicited heteronomous control is bullying, it is necessarily alien and hostile to the original, autonomous, bottom-up programmes that form the underlying structure of the hominin mind. So, it is inevitable that 'self' becomes 'self-consciously' aware of itself, being forced to find ways to enact the instructions that originated in the mind of another. As these heteronomous inputs are planted in the controlee's brain they create a platform of externally sourced intrusions, which, because they are so closely linked to the threat of punishment, virtually always gain priority over any original bottom-up programme.

This platform modifies the hominin mind by allowing

externally sourced behaviour to be registered, understood and enacted by the controlee as if it were her/his own. Being able to enact the directions of the controller is an extraordinary feat of mental gymnastics, but in addition, a remarkable quirk (inversion) appears in the hominin mind at this point that enables the controlee to *take responsibility* for these heteronomous actions, thus increasing the efficiency of the heteronomous process, and thereby reducing the potential threat of the controller. Or, couched in different terms, it can be said that this allows the controlee to become 'parasitized' by the controller; the controlee becomes subject to mental slavery, host to the directions of others – we have become puppets whose strings are pulled by others.

It seems probable that any attempts by us to experience the mindset of an autonomous animal has become more difficult because of the adaptations that have evolved in the human mind to deal with compliance to heteronomous controls. That is, there has been a selection for being 'receptive' to heteronomy; the species had no other choice. However, it seems a few religious adepts do experience enlightenment, which may indicate that a return is possible.

14:2 Personal responsibility for behaviour (accountability and imputability):

Top-down control allows the controlee to construct and enact the heteronomous behaviour that is imposed by the controller but this 'enaction' has consequences, for the controlee now becomes 'answerable', responsible and accountable for the outcome of that behaviour (because she/he is punished if it does not meet the controller's requirements). It is in being answerable that the controlee learns to modify behaviour according to the wishes, desires or needs of the controllers, whoever they may be.

The degree to which the hominin accepts top-down responsibility for the organization of its own behaviour creates a fundamental difference between mankind and the animals. This concept of top-down responsibility is difficult to grasp and articulate; it is accepted by humans as being completely normal and so it remains unexplored and unchallenged, yet it is unknown in the animal. The animal mind is not tasked to create its own behaviour nor required to take responsibility for that behaviour because all its responses are supplied to it by means of its bottom-up programmes.

Accountability for behaviour arises from the fact that heteronomous behaviour has to be enacted in a way that is approved and directed by the controller, and so the controlee becomes answerable to the controller (and to itself) for her/his own actions. The imperative to accept responsibility and behave 'properly' (in accordance with the controller's wishes) comes from the fact that the controller can inflict a range of punishments, some being extremely harsh. Remarkably, this means that the responsibility is not connected to our own genuinely autonomous actions but to the actions we enact in compliance with the heteronomous directions of others. Therefore, we become responsible for our heteronomous acts because of our own compliance, for compliance means tailoring behaviour to the controller's wishes, that is, getting it 'right' for fear of the consequences. The controller(s), however well meaning, are always 'corrupt' (in the sense that they have overridden another's autonomy) and our compliance, although understandable, is nevertheless (at the deepest philosophical level) shameful because it is achieved at the expense of autonomy.

This difference lies at the heart of human evolution; we proudly assert that it is having a greater intelligence that sets us apart from the animals, but a more startling divergence is our ability to accept top-down responsibility for heteronomously sourced activity. Being responsible for finding a response that will satisfy the controller(s), yet being the least compromising for the controlee, is an extremely difficult and uncertain process and at base absurd.

In contrast, an animal's behaviour is created by its bottom-up programmes, so although the animal is conscious of the fact that it enacts the activity, it is not responsible for constructing the behaviour it enacts, nor is it responsible for the behaviour as it is enacted, nor is it responsible for the consequences of the behaviour once it is enacted.

An animal enacts species–specific behaviour that has been selected over evolutionary time; it is not the individual's own self-constructed behaviour and so the individual cannot be held 'responsible' for these inherited behavioural programmes that were laid down millions of years before it was born. This has very important consequences; there is no difficulty, vulnerability, problem or doubt associated with triggering and enacting bottom-up behaviour; indeed, it is notable for its lack of responsibility. There is nothing in the way an animal behaves or functions that requires, needs or allows the concept of responsibility to enter its awareness, so they do not accept fault, say sorry, rectify their behaviour, or agree to act differently in the future.

So, if there is no area of the animal mind that creates behaviour from scratch (that is, works out what to do and how to do it), and if there is no area of the animal mind that accepts responsibility for its autonomous actions once they are made, and if there is no area of the animal mind that then worries about the consequences, or 'correctness' of triggering those actions – then it therefore means that there is no area of the animal mind that is tasked to monitor itself with a view to improving its behaviour.

The part of the animal mind that registers 'presence' simply floats upon whichever bottom-up behavioural programme is being triggered at any one time, it has no need to question the suitability, efficacy or efficiency of what it is doing and there is no area of the animal mind that doubts the ability of the bottom-up programmes to protect the individual from virtually every danger it may meet.

In this sense, the animal can be said to have 'faith in nature' and this is the stance (outlook or comportment) that the conscious animal appears to have when it looks outwards towards the world. The animal is relaxed as it watches and monitors the world as it goes by; it registers its own presence and its bottom-up reactions to the world, but it does not have (nor does it need) the ability, or wish, to intervene top-down to improve upon its own responses or create new responses, or set about making improvements to itself or to the environment.

With the animal, actions and reactions happen, they are triggered autonomically, so the conscious element of self is not tasked with the top-down responsibility of finding, creating, initiating or enacting any part of its behavioural repertoire. So it can be said that animals are not accountable to themselves, or to any other animal, whereas humans are accountable to themselves and to many (potentially any) other human beings.

There are odd occasions when bottom-up animal behaviour does not run smoothly when, say, a challenge to dominance results in failure. The challenger becomes anxious, nervous and apparently self-conscious of its poor judgement, but this is not 'responsibility' for its actions in the sense under discussion (getting it right by the standards of another). The challenger autonomously made the decision to challenge; it failed on this occasion and it must accept the consequences but the prize (say, to become leader of the pack) was worth the calculated risk.

Presumably this rudimentary ability of the animal to 'impute incorrectness' evolved and extended until the hominins, who are now not only beholden to the controller (the victor), they have been forced to develop a 'stance to self' where they are fully responsible for their own actions. To the extent that, for example, if pupils fight on the school bus they must take responsibility not only for their actions but for the disrepute that it brings upon their school, and in rectifying their misdemeanour they must acknowledge their imputability, accept punishment, feel shame, apologise and promise to 'never do it again'.

In accepting a remarkable level of responsibility for their actions (that is, by becoming imputable) humans have created a new pattern of behaviour which is unknown anywhere else in nature. Once 'esponsibility is accepted by the controlees then, potentially, most actions are imputable in the sense that they could have been improved and enacted differently. Christians see imputability as 'moral responsibility for one's human actions', but this ignores the fact that most human actions are under heteronomous control, so what they really mean here is that you have a moral responsibility to do what the controllers require – if only the early Christians had been able to express themselves in terms of autonomy, rather than heteronomy, we would have made much faster progress in understanding these matters.

Before the controlee can enact any heteronomous behaviour the desires, wishes, instructions of the controller need to be converted into the controlee's own actions. That is, the controlee must author movements that she/he believes will meet the controller's demands and requirements. However, by acting in this way authorship of the act becomes confused.

The action, of the physical act, is undoubtedly created (put together) by the controlee who moves her/his own limbs, but the primary cause of the act originated in the mind of the controller for he/she told the controlee what she/he had to do. So, the controlee enacts the physical act but is not the author of the concept that originated the act. Yet it is the enactor of the act that becomes responsible for the act because she/he had (was heteronomously pressured) to make the decision to enact it, and is seen to enact it, even though it would not have been created, let alone enacted, without the controller's initial coercive directions.

And because heteronomous acts are often enacted well away from the presence of the controller (albeit with the threat of later physical punishment hanging over the whole proceedings) it is even more certain that the controlee must falsely accept that she/he is responsible for the behaviour that is enacted. Clearly the controlee authors the actions that she/he enacts but given the layers of complexity there is no simple way to determine the percentage of the controlee's movements that are heteronomous or the degree to which she/he is genuinely content to create these actions.

This complexity bedevils all attempts to understand human motivation as it is impossible to start from an autonomous base to see what may, or may not, be the individual's genuine 'own' behaviour. Nevertheless, without an understanding of heteronomy, the controlee must accept total responsibility for all her/his actions; there is no alternative.

Top-down decisions have enormous flexibility, the controller only has to say, "I have changed my mind, don't do that, come over here and do this," or "Not like that, do it like this," and the controlee must modify her/his behaviour to adapt to the controller's changeable wishes.

In being able to accomplish this remarkable (damaging) feat the controlee locks her/himself into a life in which heteronomous (rather than autonomous) control becomes embedded in the mind and governs behaviour. It is the depth of this embedment that has made heteronomy so difficult to understand, for example, it is only recently that we have come to accept that in a case of rape the victim is not at fault and has no responsibility whatsoever for the aggressive invasion of the attacker. Similarly, the child that must compulsorily attend school develops a personal responsibility for her/his behaviour, performance and attendance even though it has been forced upon the child by coercive adults. We have been blind to the truth of the dynamics (power) that determine this violation because the loss of status of the victim has meant that her/his complaints, pleas and narrative are not just ignored and dismissed but they remain unregistered, which (conveniently for the controllers) means that the injustice is ignored. The aggressor, emboldened by the success of the attack, writes the history and thereby has little difficulty in getting his/her story believed – does anyone listen to the child when it complains that it has to compulsorily attend school? Against this background, to try to establish the extent of the heteronomous controls in any given society and assess the pain and damage they cause is a near impossible task; controllers have always set the agenda.

Today, we have not only come to falsely accept that we are the top-down authors of our own behaviour but that we are also answerable (accept responsibility) for that behaviour; this represents a truly remarkable and revolutionary change in evolutionary history.

We have become 'authors of our acts' to such an extent that even our deepest body functions such as defecating, urinating and mating are no longer autonomic body programmes that are simply enacted, they are now 'authored' and 'authorized' by the self to the extent that the individual is responsible for the size of the turd or the smell of the fart. No other animal is responsible for, or embarrassed by, their own excrement, genitalia, or behaviour, they simply eat, shit, piss, mate and enact their bottom-up programmes. Animals have a genuine ipseity that cannot be imputed thus the animal always remains serenely within its own proper nature. The willingness of the human to register and accept any imputation is born out of its vulnerability. If self cannot be defended from motor-control it will always be looking for ways to adjust behaviour to find and occupy a safer position.

In becoming imputable, humans opened a major difference between themselves and the animals. If you say to a neighbour, "You're a loser," or "You look a mess," "You do not look as beautiful as you used to," or "That is a very small penis," it will hit a hot spot of vulnerability in their mind and provoke an instant reaction; the recipient will be belittled and hurt, and the vulnerability will be registered and remembered for a long time. Whereas, if you say to a badger, "You are an utter disgrace, look at the state of your tail," it would just continue its journey down to the river. There is no part of the animal brain that puts its hands up and 'says', "I should have done better," or "I should not have done that," or "I wish I looked more attractive." The animal has no need to be able to manipulate theoretical, alternative, behavioural activity in this way because it is not vulnerable to motor-control and therefore has not become imputable. Being imputable creates a vulnerability to almost any criticism, which means the controlee accepts responsibility for the act for which she/he is imputed.

One's ipseity is undermined by the demands of heteronomous control, self is no longer robust, its strength is hijacked by the heteronomous controls and punishments that it is forced to endure. Self feels it is not safe (or good enough) so it is necessary to keep a lookout for any (all) so-called weaknesses of self and if possible conceal them. Note this is not necessarily because the controlee is inherently uncertain about how it wants to behave but the controller can, on a whim, find that behaviour incorrect, wrong, offensive or not leading to success and can insist upon corrections and adjustments that means that the controlee has to be able to retrace events to be able to correct past actions (errors), make amends and change behaviour in the future.

Imputability arises out of a fear of punishment, ostracism or reprisal, which means that without motor and heteronomous control, accountability could not have become an established human trait.

Humans are the only species on the planet to take responsibility for their own actions so it is bizarre that we accept this heavy burden without question, for it is clearly absurd to be accountable for unsolicited directions or standards that others have coercively planted in one's mind (often at a very young age).

The human ability to accept responsibility for a heteronomously imposed action is simply a necessity of survival. For the animal to accept responsibility in this way would be an act of utter stupidity, as it is a violation of autonomy, yet for the human it is thought to be a central pillar of moral behaviour.

The moment when an individual hominin accepted responsibility for her/his behaviour was a crucial milestone in hominin evolution. It indicated that top-down mental activity can act as a 'gatekeeper' (a barrier) that in effect says, "because I will be held accountable for my action I have to be certain that I am doing this in the best possible way, that is, correctly in the eyes of my controllers".

All inputs and outputs must pass through this template of compliance to be checked, balanced and weighed against the fear of being held responsible and punished. The top-down decisions of 'what to do', 'what not to do', and 'how to do it', are made here before the behaviour is enacted. This checkpoint has become associated with the super-intelligent, self-conscious executive function of the human self at the pinnacle of evolution but, in fact, it is the only way to function when the mind is under the repressive restraints of heteronomy. It is a mode of operation that has no purpose if the individual is not at risk from motor-control.

14:3 Duplicity:

Humans have a long history of intraspecific atrocities – murder, rape, abduction, enslavement, war, torture, lies and other violations. While these cruelties are ultimately due to the human's vulnerability to the act of overt motor-control there is an additional twist of duplicity operating here that needs closer examination.

Duplicity results from the ambivalence that is inherent in the process of enacting heteronomous controls, it arises the moment that the controlee agrees (for reasons of self-preservation) to be compliant. The controlee must be able to construct behaviour top-down that will be approved by the controller, in virtually all cases that behaviour, ultimately, will be against the inner nature of the controlee's own self, simply because it is not sourced bottom-up.

Constructing compliant behaviour places the controlee in a strange position because she/he implements and accepts responsibility for these acts even though she/he is not the original author. They are someone else's wishes, desires, instructions, which the controlee enacts, and in so doing the controlee becomes duplicitous to itself and thereby loses integrity.

There is a second element of duplicity that arises because within every individual there is a dichotomy, which means that, at different times and depending on the circumstances, she/he can become a controller or a controlee. This means that the 'controlling self' understands the distress of being controlled and the 'controlled self' understands the damage that is inflicted when she/he enforces controls on others. This dichotomy and its resulting conflicts are part of our human personality; it remains as an unresolved expression of our underlying duplicity. The self-sufficient individual animal acts autonomously at all times – it neither controls, nor is it controlled.

All of this has serious consequences for our psychological health, yet we are unaware of it because we are unable to face up to the damage that it does (and has done) to the autonomous mind. This trick or 'sleight of hand' perpetrated against oneself is repeated until it is believed, that is, self deceives itself about the source of its own motivation.

No wonder then that pharaohs, emperors, kings, religious leaders, popes, presidents, dictators, and today, the CIA and corporate business can, in their relentless drive for wealth, power and security, readily perpetrate tricks, lies and hoaxes upon a gullible populace. Ordinary citizens, already having accepted their own self-deception, have a predisposition to fall for any lie, 'psych ops' or sales pitch, particularly those generated as part of the stream of disinformation generated by the more powerful controllers. The grander, the more outrageous, the more likely the deceit will be believed, "God up in the sky is watching you", "there is an afterlife in heaven", "the Pope is God's special representative on earth", "one religion is superior to all others", "royalty are a breed set apart to be financed by their subjects", "the alphabet has phonetic rather than semantic meaning", "racial and gender superiority is natural", "a lone gunman shot President Kennedy", "man walked on the Moon", "poorly equipped foreign terrorists perpetrated the sophisticated 9/11 attacks on New York", or "pharmaceutical companies can be trusted even though they do not disclose their full trial results", "the government's sole function is to represent and protect the electorate (rather than a financial elite)". All these doubts have occupied our top-down minds in recent years. (Or, was the Emperor wearing new clothes and the young boy deliberately created a conspiracy theory to hoax the crowd with an illusion of nudity?)

Believing some of these fantasies is relatively harmless but others have led to invasion, warfare, the destruction of countries and the ruination of countless numbers of innocent families.

Greed and lack of empathy are the hallmarks of the tendency that readily disregards the plight of others, a trait that leads ultimately to the defilement of the victim and the corruption of the perpetrators. Autonomous animals are protected from this dystopia because they do not allow heteronomous propaganda or controls to enter their minds. Alas, whatever a controller chooses to say (and repeats often and forcibly enough) must be accepted by the human controlee, on fear of punishment or exclusion from the group, for there is no filter of truth or accuracy operating here, only the desires, wishes or whims of the controller.

Humans have always been vulnerable to mental penetration (our minds are as vulnerable to verbal penetration as the female body is to sexual penetration). So, if those in power wish us to believe 'x' or 'y' it is relatively easy (with money, publicity, the production of a spectacle and today the control of the media) for them to achieve that end. Here is the dark art of social control for it is now possible to manipulate and control the viewer with psychological operations so that physical (true) facts are denied even when they were plainly visible and can be viewed again and again on a YouTube video. By having a safety lock that bars heteronomous intrusions into their mind, animals are spared a lifetime of falsehood, manipulation and degradation, for unlike the human they can believe and trust the sensory information that enters via their own eyes and ears.

It is extraordinarily difficult (if not impossible) for the individual to maintain a healthy vitality in the knowledge that she/he has lost autonomy in the past, is vulnerable to further losses in the future and at present is heteronomously acting out the desires, wishes, and directions of others. Indeed, some find it impossible to function burdened by their compliance in their heteronomous predicament and as a result become ill or suicidal.

The need to maintain conatus is so essential that in all other forms of parasitism (heteronomy is essentially a form of parasitism) the host proceeds with its life unaware of the burden that the parasite imposes. By repressing, with the aid of alcohol, drugs, entertainment and holidays, or simply deceiving self about the essence of heteronomy (that it is imposed from outside), self can ignore the fact that others have entered its mind and are able to control its behaviour. We choose to deceive ourselves because that is much easier than accepting the demeaning truth of our own participation in our own enslavement.

The top-down hominin brain has been selected for its ability to regulate a very difficult situation, having to balance the demands of heteronomous controls with its own essential autonomous requirements. Given the complexities of this task (in which we are all engaged) there is no easy way for us to understand (conjecture) what it would be like to have an autonomous brain that functions bottom-up.

Top-down control is born out of serious difficulty for self is no longer a self-sufficient autonomous self; it has become a differentiated self where autonomous needs must be fitted in between periods of heteronomous control. But, as this is the only self that we experience, we take it that it is the 'real' self (or 'the only possible self') but it is a duplicitous imitation of the truly autonomous self that lies deep within us. The human self becomes divided, treacherous and false from the moment it complies with the first heteronomous top-down control. This is the nature of the human condition; our 'stance' to the world reflects the tragic fact that we are no longer fully autonomous beings. We implement the heteronomous controls that have been planted in our mind and as we grow we supplement the close family controls with controls that come from teachers, friends, employers, the state and others in society. We have evolved to be able to mould ourselves to the requirements of any controller (that is the nature of heteronomous control) and in so doing we have lost the integrity and equanimity (wholeness, cohesion, unity) of the autonomous animal.

We need to face the truth about ourselves; the perplexing boundary between animal and man arises because we have all become victims of heteronomous controls. This indigestible, disconcerting, yet ultimately transformational, conclusion is developed in the next chapter.

CHAPTER FIFTEEN

BEYOND HETERONOMY

As the hominins gained the ability to motor-control we became a danger to each other and the landscape of our existence changed. It was no longer possible for the individual to look out onto the world and be confident that her/his own bottom-up programmes were adequate for survival.

A dystopian continent had opened that heretofore had never existed, it is a land that lies beyond the point of being held against one's will, past the point of submission, where the fearful individual must comply with directions of others, construct her/his behavioural responses and accept responsibility for their enaction.

The moment the controlee is held, controlled and directed, her/his autonomous existence comes to an end. Survival beyond capture opens a harsh new world where hominin life is possible but only in a heteronomous form. This is the world in which we all now live, for everything that humans have created including their artefacts, relationships and thought processes are the products of heteronomy. It is an area of nature that no other species has lived in before and, so it remains a territory completely outside of animal experience. For as Rainer Maria Rilke said, "Never, not for a single day, do we have before us that pure space into which flowers endlessly open."

Having to survive in this hostile environment necessitated the development of a top-down organization of the mind which enabled the close interpersonal compliance and cooperation that is integral to the maintenance of a heteronomous relationship. From an early age the individual is forced to respond, cooperate and adapt to a continuous barrage of coercive, heteronomous controls.

External controls do enormous damage because they inescapably override autonomy, destroy confidence, shatter equanimity and in exchange we are left with fear and uncertainty; humans have always been a danger to their own species. The animal mind is not charged with the necessity of constructing its own defences and being responsible for them; it simply enacts defences that are provided bottom-up by nature. Therefore, the animal mind exists in a fundamentally different type of space to that of the human mind. The animal mind floats in an accepting, non-responsible, and non-questioning space that is radically different to the threatening, uncertain and troubled space within which the human mind struggles to comply with the directions and expectations of others.

Human beings experience agitation, anxiety and top-down control (and inflict their problems on others) whereas the animal experiences hereness, equanimity and acceptance and it relies without question (faith in nature) solely upon the bottom-up programmes which it inherits. These bottom-up programmes, having been selected by evolution over millions of years, are the most effective and reliable that can be constructed, so (in a world without motor-control) it would be pure folly for the individual animal to think about replacing them with its own short-term, improvised, top-down responses.

We need to see that humans have been ensnared by motor-control, whereas animals have not, and to understand just what it is that we have lost and what animals have retained. We have lost our place within the natural animal world (paradise) that operates upon the principles of individual autonomy and replaced it with a dystopic world of heteronomous control. This change, probably the most fundamental in animal evolutionary history, results from the titanic (but natural) clash between the male drive to reproduce and the female's absolute necessity to establish anoestrus.

It is a tragic confrontation that became inevitable the moment hominin hands were free to grasp and motor-control. There was no element of choice here; the hominins lost their safety (and their equanimity) because of unique but natural circumstances. It was not given up or taken away due to carelessness, or forfeited as some sort of punishment (certainly not for Eve's disobedience), it was lost because the hands and arms of the hominins were freed from their original function in the trees. But it did have the most catastrophic consequences and the hominins fell into an existence of incessant coercion and heteronomy, which is still our daily experience. The painful nature of these difficulties has left us with a deep-seated denial of our predicament, for to face up to the horrors of the present situation would involve a recognition that the exceptional element of human exceptionalism consists in us being the only species that has fallen victim to heteronomous control.

Before the hominins used their hands to motor-control there had never been a reason (or a method) to get outside of bottom-up life. By living autonomously, the animal is embraced within the 'ongoing flow of life', it is unable (it does not want, or need) to get outside this flow and so it does not (cannot) consider its own behaviour from an outside perspective; the animal has no need for a cognitive revolution for it accepts all that is given.

This has been the *modus operandi* of all motile creatures for hundreds of millions of years and no selection pressure had arisen (before motor-control) to change this fundamental relationship between the autonomous animal and the outside world. For the first time in evolutionary history, the individual early hominin was left in a situation where, to survive heteronomous controls, she/he had to take personal responsibility for constructing their own top-down responses. This overturned the order of existence for when the process was eventually completed we were no longer autonomous animals within nature, we were heteronomous hominins outside of nature. We had developed a brain capable of responding to the directions of others, we had become puppets (mental slaves) at the behest of any individual, or group, powerful enough to control our behaviour.

This is a widespread defilement and degradation, from which the animals have been fortunate to have been spared. It is not just that we can be held by another, or raped or tortured by another, sold into slavery or held in servitude by another, or that the mind can be penetrated, deceived, abused and made to act upon the dictates of the heteronomous controls of another – it is that we experience and live in fear of these things everyday of our life whereas the wild animal does not. Every act of motor-control is a loss of autonomy that waits to be redeemed.

So here is the central question for the philosophy of evolution: if we have evolved to respond to the heteronomous controls forced upon us by others, does that mean we are at the apex of evolution, more intelligent (superior and exceptional) than the autonomous animal, or are we simply exhibiting the symptoms of a tragic vulnerability to motor-control?

With the advent of motor-control the essence of animal existence (which is freedom, autonomy, life in the open, hereness, equanimity, and no top-down responsibility whatsoever) was moved so far out of the hominin reach that it effectively disappeared (became unavailable); whether it can be regained is a question for the future.

Being forced to live in a heteronomous world has blinded us in so many ways but three have had major repercussions. First, although we claim to be exceptional, we have little or no understanding of what it is like to live in the animal world of autonomy and equanimity because it is not something we have ever experienced. Second, we do not understand that we have been excluded from a world where autonomy is paramount (paradise) and thirdly, we do not understand that paradise is a state of existence that humans could, in theory, endeavour to regain.

From the moment hands were used to motor-control, humans have been on an unprecedented journey. So, let us return to the musings of my earlier daydreams (Chapter 1) and the moment it all began. I suggest that if the original anoestrous protohominin female (who was chased by the males on the African plains) had somehow been able to reliably avoid capture and make her escape then hominin evolution, as we know it, would not have taken place. However, this outcome would have necessitated an inability (or reliable reluctance) on the part of the male to use his hands to grip and hold the female and it is difficult to see how this trait could have evolved once dexterity had been selected in the treetops. Today, theoretically, it would be possible to immobilize the thumb at birth, but this would be an extreme act of heteronomous control, far from an expression of autonomy. Hence the conflict between the male ability to hold with his hands and female's need to establish anoestrus remains tragically unresolved.

As mentioned in the last chapter, a controller and a controlee lie within each of us and so we have ambivalent thoughts about the best way to come to terms with this situation. Depending on size, power and intent, we are all capable of controlling others and we are all vulnerable to being controlled and we can change from being the controller to the controlee in a moment.

We need to learn not to trespass upon one another just as we have learnt not to poison our own drinking water, or set fire to the village, or foul the bed, or destroy the food crops, or slaughter the breeding cattle. If these things can become second nature then so can an abhorrence or disgust of one human holding and controlling another, for as the incidence of motor-control escalates, society is increasingly losing its connection to its autonomous roots.

Once heteronomous control overrides autonomy then a dystopian world of social power is created, heteronomy is power over others. Autonomy is self-governance, it is not power over others, it is power over self. Autonomy cannot exert power over others unless it can catch hold and motor-control.

In the animal world, without motor-control, autonomy operates autonomously, conspecifics cannot be caught (or are deliberately not caught) and as a result their social life has an underlying long-term safety and stability. I suggest that the reason we all have difficulty seeing the damage caused by heteronomous controls is that we are so completely ensnared by them, and so they appear to be the only behaviour that is open to us. Having no alternative, we succumb to the overwhelming power of heteronomy (even claiming that it represents the pinnacle of evolution) not understanding that animals have never had to adopt anything like this dire, demeaning, top-down behaviour because they have never been parted from the perfection of their own (bottom-up) autonomous governance.

The present temporary solution to all of this (in ignorance of the operation of motor-control) has been to draw up social contracts of one sort or another that seek to ameliorate the worst excesses of motor-control. Social contracts can never return full autonomy to the controlee because the ability to hold and control remains an inherent threat. The controllers will always make sure that they have the strength, power, wealth or prestige to manipulate the contract to their advantage. This dilemma is at the heart of all political and economic discourse. For autonomy to be fully restored to all members of society, the controllers would have to (autonomously) self-restrict their ability to control conspecifics. If the controller is restricted by any means other than her/his own autonomous control then she/he immediately becomes a controlee, controlled by others, which thereby increases (rather than diminishes) the sum of heteronomy. Democracy does not resolve this dilemma, no animal votes to be governed by others. Why should they, they are already freely governing themselves?

Here is the deception and the dilemma at the heart of human behaviour, for up until the evolution of the hominins, autonomy had always been associated with the 'perfection' and freedom of expression of our individual bottom-up programmes but now it leads (via the autonomous lustful male) to motor-control, heteronomy, coercion, rape and sin – a truly diabolical outcome.

This is how evil has come to exist in a universe created by a benevolent nature (or a loving God). Given that hands evolved naturally to allow monkeys and apes to live in arboreal habitats they cannot be classed as an error. The 'imperfection' here (if that is the right word) was the danger of retaining an organ that could grasp, grip and hold after the arboreal/terrestrial transition when it was no longer required for its primary purpose of holding onto branches in the trees. But evolution is not (cannot) be aware of future consequences of the changes it selects, these can only be expressed as later survival rates.

In surviving motor-control (owing to the 'gentle' use of hands) hominins have unleashed a battle between the autonomy of the controller who seeks to control others and the autonomy of the controlee who wishes to be free. This is a conflict of epic proportions which is considerably more destructive to social cohesion than other great antagonists such as life against death. The difference between surviving motor-control or succumbing to predation is that with predation the lost autonomy of the consumed prey does not linger on and disrupt society. Whereas when the hominin experiences motor-control, the controlee survives, but her/his bottom-up governance is so impaired that the species' social cohesion is inevitably disrupted. A conflict is set in motion in which the compromised autonomy of the controlee is locked into a lifelong battle with the trespassing autonomy of the controller (and all of those who assist the controller).

Several questions arise from this antagonism: do we as controlees do nothing about our plight and simply continue to live as puppets under heteronomous control? And as controllers, do we continue to control and override the autonomy of others because we have such a deep-seated fear of being controlled that we control them before they control us?

Or, do we face the truth about ourselves, come to understand the biological severity of the problem that motor-control creates and endeavour to re-establish autonomy for everyone? If that were possible we would then regain the status of the animal able to protect autonomy with ease.

It remains to be seen whether we will be able to unravel this complex situation; it is possible that human intelligence is too intimately linked with heteronomous controls to be able to sufficiently empathise with the controlee and to concede the principle that everyone has a right to have their full autonomy restored, respected and guaranteed.

Or perhaps this is too pessimistic. As controllers, may we come to see the benefit of relinquishing our ability to heteronomously control? It appears counter-intuitive that we would be willing to concede power in this way, but it must be remembered that all controllers will, at some stage, have been a controlee subject to the heteronomous controls of parents, family, schools, state or enemies. Also, our ability to control is now creating insuperable problems for the planet, so it is possible that we will see that safety arises from an inhibition of control (as in the animal) rather than the enforcement of controls (as in the human), thereby allowing autonomy the freedom to regain its full expression.

15:1 Where science and religion merge:

There are other questions that are relevant to this new biological assessment of humankind's place in nature, which appear to fall outside of scientific explanations but on closer examination are seen to have a rational grounding. These include questions such as, why does religion continue to have relevance for so many people; to what are the religious referring when they speak of the sacred; and why is science still unable to handle this subject?

Obviously, animals exist within nature; they have evolved over hundreds of millions of years and throughout that time they have had no need for religion in any form. Whereas humans (as we know them) have been extant for less than one hundred thousand years, yet we have the audacity to claim a relationship with an unexplainable God that is allegedly more special than the relationship that the animal has to the natural world.

The wild animals, living within the open expanse of nature, can clearly be seen to have a presence, a 'sacred aura' about them (ask any naturalist) and animals (across all species) exhibit a noteworthy equanimity which appears remarkably like the state of bliss that is sought by all the major religions. In contrast, humans are ill at ease, we are distressed, discontent and anxious and our group behaviour is so appalling at times, it is clear we are in urgent need of salvation.

Lacking any understanding of the effects that motor-control has had upon human behaviour it has proved impossible to formulate these deeper religious thoughts in meaningful biological terms.

One of the problems relates to a major miss-move regarding autonomy that stems from the Abrahamic religion's insistence that heteronomous control was (is) a force for good, in that it facilitates obedience to the words of God (which has the corollary that autonomy is sinful).

The *Modern Catholic Dictionary* says under the entry for 'heteronomy' – "The moral doctrine that man is subject to laws that are not created by himself, but by others, and ultimately by God." (The opposite of autonomy.) Interestingly the above dictionary has no entry for 'autonomy', however, the *New Dictionary of Christian Theology* says – "Other Christians say that religious knowledge is autonomous, rather than being based on an external given revelation of God (heteronomy). But *autonomy* can be a proud human self-reliance totally at odds with proper dependence and obedience to God. *In this sense autonomy is a primary manifestation of sin.*" (My italics.)

Here is the genesis of the idea that autonomy must be overcome because heteronomy (being subject to the laws of God or his representatives) is the desired state. This confusion arises because we fail to understand that, in operating via the top-down neural pathways, heteronomy functions virtually independently of the bottom-up programmes (the original laws of Nature/God). This is how humans have come to occupy their strange position of being somehow outside nature while at the same time being totally dependent upon it. Heteronomy, in evolutionary terms, is a very recent miss-move (there was barely a hint of heteronomous control in the world before the hominins descended from the trees).

Heteronomy is thought (in this Christian definition) to be desirable because it keeps autonomy in check. Autonomy is thought to be undesirable because, in the final analysis, it leads to sin, particularly unrestrained sexual coercion. Thinking this way allows the controllers to continue controlling – even though it is the controller's controls that caused the problem in the first place. The controlee has an urgent need to restore her/his full autonomy but if autonomy is encouraged (rather than restrained) it carries the implication that the controllers (in keeping with their autonomous desires) will go on controlling, leaving the controlee in an untenable position.

The key to breaking out of this bind is to understand that the autonomous bottom-up switching mechanism was honed to perfection up until the moment it started to motor-control conspecifics. The moment an autonomous life force gains advantage for itself, at the expense of a conspecific's autonomous life force, then insuperable problems are created for the functioning of hominin society. If this is to be reversed so that autonomy is restored, all physical and heteronomous controls will need to be abhorred by the entire community.

At base, mankind is as subject to the bottom-up laws of nature as are the animals; these laws are not created by humans, they arise from nature and they are expressed through autonomous (not heteronomous) governance. Nature, God and autonomy are one in this respect, so biological obedience to autonomy is obedience to Nature's laws (or God's laws), whereas obedience to heteronomous control is obedience to human laws (an obedience that is inherently blasphemous).

Animals do not have the need to get outside of nature to question whether they should modify their bottom-up programmes (and without motor-control neither would we). Nor do they have the need to question any of nature's biological principals (and neither should we) for they should simply be accepted as given. Humans are destroying (rather than living in harmony with) their environment, which is an indication of our psychopathology and of how estranged we have become from the ecosystems from which we have evolved.

In the past, Christianity was riven by disagreements about the nature of creation. In disagreeing with Gnosticism, Bishop Irenaeus of Lyons (125–202), said that there are not two separate and antagonistic creations, but a single creation in which human beings advance towards God and salvation.

However, if these first Christians had taken note of earlier religions (as Nietzsche suggested in *Thus Spake Zarathustra*) they would have better understood the nature of this division – for the dualist Zoroastrians (and later the Manichaeists) had already clearly stated that the battle between darkness and light was of long but *limited* duration. There being three phases, the first, of peace and righteousness, before the antagonists are created (the time before the hominins introduced and succumbed to motor-control), the second, in which (present-day) mankind is intimately involved, when the antagonists come into conflict, and the third phase when the battle is resolved, with man's assistance, and the antagonists are parted, and righteousness and peace are restored.

Parting the antagonists is as simple as not using our hands to motor-control others but achieving this outcome will not be as simple as stating the principle; nevertheless, understanding the predicament is a step towards resolution. The conflict between autonomy and control (dualism) will only be overcome when we understand that the controller and the controlee have become antagonists rather than sociable conspecifics. When the controller overrides the autonomy of the controlee it violates the principle of autonomous governance upon which all animal existence is based, for it is biologically unsustainable (despite the recent – but temporary – human population explosion to the contrary) to attack, restrain, and control conspecifics in this way.

Two thousand five hundred years ago, Zoroaster must have realized all this, for he said that, "the antagonists would eventually be parted, and the battle resolved." These major questions have been discussed by the major religions throughout history and we owe them respect for their pioneering thoughts because they do carry the seeds of a much-needed fundamental reappraisal of the human condition.

If the 'tree of life' in the Christian scriptures is understood as representing autonomy and the 'tree of knowledge of good and evil' is understood as representing heteronomy, and the very tempting apple (sexual pleasure), which is harvested by the grasping hand, is understood as representing the plunder of motor-control – then we have a way to unravel the enigmatic passage in Genesis 3 verses 1–6.

We can also see that there is nothing supernatural here but there is a major complication, for embedded in all of this is a deliberate and appalling miss-move. It is said in this parable that disobedient Eve took the first bite of the apple which led to the downfall of humankind but clearly it must have been Adam's 'disobedience', for the original violation of autonomy could only have appeared because of unleashing the male sexual drive. The fruit that has been plundered is the anoestrous female. If Adam had delayed (shown restraint) until the female was in oestrus, then the 'tree of the knowledge of good and evil' would not have taken root. It is not sexuality *per se* that is the problem, it is the violation of the anoestrous female that is sinful (unjust, unrighteous), for it results in a catastrophic loss of female autonomy. The female has always been (always will be) tempting to the male, that is an essential fact of animal reproduction, to blame her for her own loss of autonomy is to victimize her for being a victim, truly the work of a wily male serpent.

While mentioning these religious themes, it is worth noting that St Augustine's concept of original sin, passed down the generations via the seed of the male, turns out not to be a genetic problem but a classic behavioural, environmental problem like the presence of cholera in the water supply. If you stop drinking the polluted water, the disease outbreak is contained and health returns. Similarly, if you stop perpetrating acts of motor-control then heteronomy, coercion and compliance (which can be understood as a disease resulting from original sin of motor-control) are contained and autonomy, health, freedom and equanimity return spontaneously.

The crucifixion is a perfect symbol of suffering and sacrifice but it can now also be thought of as the ultimate symbol of motor-control. The victim's hands are nailed to the cross so that he (she) remains alive for several days, unable to escape, immobile and defenceless to attack until death mercifully intervenes. Contrast this with the gesture of prayer where the hands are gently placed together and held in a position where they indicate they can do no harm (they cannot grasp nor, can they impose controls), not only is this a symbol of 'I shall do no harm to others', it is a sign of respect for autonomy, a recognition that freedom is a gift bestowed by nature (or God).

Gershom Scholem said (and I paraphrase Giorgio Agamben, *Potentialities*, page 163), "because we cannot decipher the scriptures we cannot return to our originary condition; hence, we need to discover and understand the form and content of the law before the Fall." That is, we need to face the truth about ourselves, we need to heed the suffering of our lost autonomy and realise that the original purpose of mankind's invention of God (in all religions) was to create a vehicle that first, comforted and supported us dealing with the horrors of heteronomous control and second, introduced the first tentative concepts to help us come to understand and regain our autonomy. It is telling that animals have no need to find God, for the 'essence of God' (which we lack and choose not to see around us) is surely present in every untrammelled movement of the autonomous animal. The law before the 'Fall' was predicated on the need to allow autonomous freedom for all living creatures, it was a time when deliberate intraspecific motor-control was unknown – neither God nor Nature had even dreamt of such a thing.

The Bible says, "In the beginning was the Word and the Word was with God and the Word was God" (John chapter 1 verse 1). And in Buddhist literature it says, "As to what stands prior to the Word, not one phrase has been handed down, even by a thousand holy ones." (*Two Zen Classics* page 386, Katsuki Sekida). It is clear we have not gone far enough back in time to seek the 'law before the Fall'. When we do so we will see that it is the evolution of animals, with their pure, silent autonomy, that preceded the Word. Animals have lived and functioned perfectly for millions of years without a single word ever passing their lips. I suggest that in the search for human enlightenment we will be led back to the pre-verbal world of the animals, where relationships are conducted autonomously without the need for the heteronomous controls that drive human language.

Wild animals, who have no attachments (apart from family), have a mental stance to the world that is very similar to the position described in Buddhist texts which says, "Emptiness is the ancient Way, before the 'ten thousand myriad objects' of the mind were born." Very young children enjoy their positive *samadhi* instinctively and so do horses and cows standing peacefully in the sun. In a state of emptiness there is no dread, no anger and no fear of death; everything directly presents itself (*Two Zen Classics*, page 40).

Unfortunately, the Buddha had nothing to say as to why we had become divorced from this ancient way. More than that, he said it was foolish to waste time searching for reasons; the task was to gain enlightenment by emptying the mind, breaking all attachments and seeking Nirvana. Wise as this is for the individual, we are left with an unanswered question – how did we become divorced from enlightenment? Had the Buddha known he surely would have told us, for it makes the quest for enlightenment that much easier if it is understood as an established feature in the natural world. A world where animals retain the enlightenment with which they are born, and humans lose their enlightenment because they are exposed to heteronomous motor-control.

The human is the only animal to live in a world of heteronomous controls, therefore any awareness, understanding or glimpse of autonomous governance would be a transcendental experience, opening a world above and beyond our present everyday heteronomous responsibilities, tensions, struggles and oppression. If you, or I, were somehow to suddenly experience the autonomous freedom of the animal it would be a revelation. It would also be a hallowed moment in the sense that, heteronomous control (the original sin) has proved to be so disruptive to hominin behaviour that any restoration of autonomy must be considered an act of a sacred significance, a moment of redemption.

Autonomy (before it gained the ability to control others) has such a fundamental role in the organization of behaviour that it naturally possesses an aura or a reverence without need of enhancement from any supernatural aspect of religion. The properties of autonomous governance are solely biological (science-based and of this world) – we are blinded to its presence by the ubiquitous duplicity of heteronomy, so when the equanimity of autonomy is experienced it appears to be sourced from another world. It is only in the full experience of autonomy that the religious dimension of human nature will be understood. The need for organized religion will simply fall away for the essence of the sacred and the divine is none other than the unfettered expression of a completely natural autonomy.

All these reassessments wait to be absorbed into the mainstream thinking of both science and religion and when that occurs we will be much better placed to decide to what degree (if at all) science and religion are antithetical.

Thus, with our secular feet planted firmly on the ground, we can now see why, in spite of the lack of any tangible human experience, the major religions have steadfastly refused to give up on the idea of the sacred, of salvation, enlightenment, awakening, serenity, bliss, the unnameable, release, liberation, transcendence, nirvana, *bodhi*, 'kenosis' (in the sense of emptying one's heteronomous self), *satori*, *kensho*, *moksha*, being, pure consciousness or something greater than we experience at present, a longing for existence outside of the experience of time.

This is already the animal experience of everyday reality (not that any religion claims it to be so), for how else could animal behaviour exhibit such equanimity in all situations and conditions. We need to consider the possibility that autonomous governance is not only the basic organizing force of motile nature but that the unfettered expression of autonomy produces an associated feeling of safety, contentment and bliss, which is the everyday experience of the animal mind, and this (I suggest) is the same mental state as the 'enlightenment' (or 'union with God') that is sought by religious adherents. The difference being that animals have not been parted from their autonomous governance, so they have no need to seek redemption (the cessation of heteronomous control) whereas, having lost autonomy, we now have an urgent need for it to be regained.

The tragic miss-move made by the Abrahamic religions (which was a triumph for the anthropological machine) was to turn their backs upon the animal, for the animal that we see before us has not left Eden. Every hour of the day the animal mind basks in a state of equanimity which can only arise from the absolute trust it has in the ability of its bottom-up programmes to always keep it safe.

We need to recognize the vast reservoir of autonomous experience that animals exhibit for without their example (guidance) it will not be possible to rediscover the autonomous behaviour that is latent within ourselves. However, before we 'return from our exile' (or before our apocatastasis sets in) we would first need to establish a post-motor-control world (or re-establish the pre-motor-control world) where autonomy was paramount. The animal exemplar offers a daily, tangible, silent witness to (disclosure of) the hidden perfection of the autonomous life – it is not something we can reach out and grasp for it is only available to those who, in the absence of motor-control, are blessed to live their life autonomously.

CHAPTER SIXTEEN

SUMMARY

This book has been written to introduce the idea that there is a fundamentally different way of 'reading' human behaviour. We have failed to be sufficiently critical of our own actions, which are bizarre compared to animal norms. We misread our own species because our thoughts about ourselves contain a major misconception that has blinded us to the truth. We are not the super-intelligent ape that we think we are; we are a species struggling at every turn to deal with the effects of falling victim to heteronomous control.

We believe we are rational and civilized and with our superior intelligence that we are well fitted to run society; this view of ourselves is never seriously challenged although it says nothing about how, or why, we arrived in this uniquely strange position. Nor do we know why we are the only mammalian species in which the females lost their ability to defend anoestrus and as a result became permanently receptive, or why we all have found it necessary to develop a top-down mode of mental organization, or why we need to be civilized, educated and to have social contracts to be able to survive.

Hence, we do not understand how we have come to find ourselves occupying this strangely dominant and destructive position in the world. We live in a bubble of heteronomous controls, but we do not know that we live in that bubble. We have little information and no experience of living in the autonomous bottom-up world inhabited by all the wild animals and no understanding that our top-down world is so anomalous.

I have attempted to think through the human condition, not from the point of view of our current grandiose beliefs of superiority but from the natural interactions out of which we have been forged. We need to see our behaviour as a symptom of a malaise from which we all suffer and if we can reread ourselves in this way then we should be able to start to focus on the malaise rather than the symptoms. Our exceptionalism results from our extreme vulnerability to motor-control and proclaiming our exceptionalism has become a cover for our inability to regain autonomy and find safety.

We do not comprehend the biological power and the scale of the disruption that motor-control has had on our species and its relevance has not been included in any framework of psychology, psychoanalysis, psychiatry, politics, education, healthcare, or religious, capitalist or socialist discourses. Whichever current theories you now hold, be prepared to see how they can be challenged and improved by the inclusion of an awareness of the effects of motor-control.

Unashamedly, this is a theory of freedom that seeks to find a way out of human bondage; it seeks to find a life that is not constantly overwhelmed by power and control. Bondage results from the use of motor-control and it came into existence the moment the first hominin was held past the point of submission. From that time hominin evolution has been driven by a constant battle to mitigate the worst effects of these species-generated controls. Could the end of that long battle now be coming into sight?

First, we need to become aware that our large opposable thumb allows the hominin hand to create a powerful grip from which it is difficult or impossible for the victim to escape. This creates a situation where being held and motor-controlled becomes a constant threat to the species and I suggest that from this intraspecific selection pressure a series of consequences unfolded that led to the divergence of humans from animal norms.

Being held and motor-controlled led to a loss of autonomy and to coercive mating, which in turn led to:

1. bipedalism
2. loss of anoestrus, loss of an overt oestrus and the emergence of permanent receptivity
3. loss of hair cover, reduction of canine teeth and the increased size of breasts on non-lactating females
4. encephalization and the top-down organization of the brain
5. loss of animal equanimity and the loss of animal enlightenment
6. introduction of the concept of being responsible for one's actions
7. development of speech and language
8. the social contract, civilization, domesticity and the wearing of clothes.

This list of specifically hominin/human features, when examined more carefully, is, I suggest, not of characteristics of which we should be proud *per se*, for they are all responses and defences to the experiences of heteronomous motor-control and we need to start to view them from this new perspective.

There are two main interconnected issues that have developed out of the loss of autonomy: first, the vulnerability of the hominin/human female to coercive mating. Motor-control led to a catastrophic loss of autonomy and we all fall under its thrall, but the female suffered grievously. Without assistance, she is no longer able to defend anoestrus, which led to her permanent receptivity and forced her to abandon her period of overt oestrus. This major vulnerability inexorably gave rise to the development of the hominin's top-down mental organization, which facilitated the new patterns of relationship that enabled the permanently receptive females to continue living alongside the males in groups, or pair bonds. This was done by allowing the male much greater sexual access to the female than he would have had as an animal (which was limited to her oestrous period only) but less than the unlimited access afforded him by means of unrestrained motor-control. This meant that the female accommodated, as best she could, an increase in the sexual demands from the male on the understanding that she would not be overwhelmed by his excessive attentions. This compromise was the first social contract and it eventually gave rise to our elaborate codes of morality that attempt to regulate the worst excesses of male–female conflict. However, in many societies today the codes of morality restrict the female to such an extent that she still suffers unconscionable constraints.

The second main issue is the loss of equanimity in the hominin mind, which inexorably developed out of the loss of autonomy. Heteronomous control differentiates the hominin mind by splitting the controlee's mind in two, one-part monitoring self, whilst the other part monitors the controller(s) to be able to comply with their demands. Part of any response always lingers as problematical for it could always have been enacted differently, and so it leaves a residue of concern as the controlee can never be certain that the controller's expectations have been met. There is no closure, one is always left with a myriad of demands and alternatives, nothing is ever finalised, closed, fully discharged or forgotten as it is with bottom-up behaviour. Hence our memories expand and our conjectures into the future multiply. Paradise cannot now be experienced because the controlee has lost the peace and security that arose from behaving autonomously and so the individual no longer experiences the world from a unitary platform. In being subject to heteronomous control we have lost the dignity of our individuality. Our actions are no longer shaped autonomously by our bottom-up programmes (selected by evolution) but by the top-down, short-term heteronomous directions of others.

Being forced to enact heteronomous controls (and at the same time develop defences to them as best one can) has led to the growth of a top-down organization of the human mind. No animal, before the hominins, has found a method to break out of its species' bottom-up behaviour to construct its own top-down responses. No autonomous animal has ever had a need to do so because they have never been subject to the invasive pressures of motor-control.

Top-down mental organization enables the controlee (under duress) to enact the wishes, desires and controls of the controller and to look for new, seemingly better, defences to try to overcome its vulnerability to motor-control. This has created an eagerness for new behaviour associated with personal gain (at the expense of others). We have become addicted to activity, newness and change; we explore new territory, make new artefacts, create new buildings, make new weapons, fashion new ornamentation and develop new communications, all to improve our defences by making something of ourselves and gaining status. However, compared to the animals, all this behaviour is a serious affliction, it is simply the most recent way our species has developed to deal with the stress of motor-control.

We need to think our way into the animal brain and see that it is impervious to heteronomous control and therefore has no need for a top-down mental function. The animal mind is waiting and prepared to enact bottom-up behaviour; it is not waiting and prepared to enact heteronomous top-down behaviour in the way that humans are, for the animal mind is blank to the need so to do. Therefore, the animal mind cannot be triggered, or sprung out of its bottom-up state. When I ask the deer in the woods, "Why don't you do any work?" there is no reply – it is an indication of their autonomy and safety that they do not need to respond; why should they?

We do not understand what the experience of being free from heteronomous control would feel like, or how it could be achieved. In principle, not one response, word or thought generated by heteronomous control should be in our minds because it is illegitimate and antithetical to an autonomous life (there is not one heteronomous thought or response in the mind of a wild animal). If heteronomous control were eliminated from our behaviour, top-down mental organization would become unnecessary and as it atrophied it would allow autonomy to regain its primacy and the individual would experience the freedom and equanimity of the animal mind.

Animals do not hold, or grip, members of their own species past the point of submission so they do not get a purchase on the body, or the mind, of another animal to manipulate its behaviour, so a potential controller is unable to intimidate others and persuade them into acting the way he/she wants. This creates an intraspecific security that animals take as given but it is a security that hominins have lost and need to regain.

Heteronomous controls have given rise to a remarkable new phenomenon – the controlees have become responsible to the controller for the 'correctness' of their actions, despite the fact those actions were sourced in the controller's mind. That is, the controlee is made to be responsible for enacting actions that in truth are not her/his own – "Have you done what I told you to do?" This creates a unique situation; the controller in effect parasitizes the mind of the controlee so that the controller's wishes are carried out using the controlee as a source of energy.

Controlees need to enact their own basic actions while at the same time find ways to enact the directions of the controller;

it is a complex mental manoeuvre to balance these conflicting demands and it can only be carried out by a brain that has been differentiated and is able to operate top-down. However, it is beset by an intrinsic problem: the controller is a separate person, so she/he cannot have an inbuilt feedback regarding the actual bodily state of the controlee, which means that the controller's demands are not self-regulated to remain in harmony with the physical and mental abilities of the controlee. It is this lack of feedback that gives rise to the massive imbalances, tensions, fear, pain and insecurity in human behaviour and ultimately denies heteronomy any legitimacy.

The anxiety arising from motor-control is a major tension which is extremely harmful not only to the individual, but to our own species, to all other species and to the planet. Before we started to coerce and control our conspecifics we were autonomous primates surviving in the natural world by means of our bottom-up responses and this is the world to which we should be seeking to return. We now believe that we can do whatever we want, we behave as if we have broken free from nature using our superior intelligence and our top-down control of the mind, but this is a very damaging, grandiose view which denies the underlying hurt and insecurity from which we are all suffering. Human exceptionalism needs a deep analysis, for belief in our superiority is central to the false understanding we have of ourselves.

There is an interesting double narrative here, for while I am arguing for a return to bottom-up functioning it is only the top-down organization of the mind that can start to resolve this dilemma. Here is the paradox: while heteronomous control is extant, top-down mental activity is essential, yet top-down mental activity is indicative of a triple loss, loss of autonomy, loss of bottom-up mental organization and a loss of the experience of paradise. Hence, we need to find ways to eliminate heteronomous controls from human behaviour so that we can recover our autonomous function.

All animals in their natural habitat are born into freedom (which they maintain up until the moment of their death) and they live autonomously in the sense that all their actions are in accordance with their species' bottom-up programmes. They are never told what to do by others, or forced to do things they do not want to do. They do not have a spoken or written language, yet they live socially. They know all they need without talking. Their minds are tranquil and equitable; there is no need for them to 'improve' their behaviour because it has been selected to be in harmony with their environmental niche. If humans forsook their motor-controlling habits they would experience a similar stance to the world.

It is the ability to hold and control that has inexorably led to the development of weapons, guns, social contracts and laws that defend the controlee on the one hand and give advantage to the controller on the other. The hominin hand was the first weapon in this series, it created an unprecedented loss of freedom for the controlee and it led to a unique system of controls to which we all fall victim. We have been unable, at any time in the species' history, to create a totally adequate defence to these controls but we have developed partial defences which have come to be understood as being representative of the uniqueness and superiority of the hominin species. These adaptations now need to be seen for what they are – the way in which we accommodated the new and threatening selection pressures that arose following the use of the hand to motor-control.

Once the hominins established a top-down mode of mental organization it virtually always gained priority because it overrode the original bottom-up (animal) organization of behaviour. Top-down mental organization also critically monitors self for weaknesses and defects (in fear of motor-control). This introduced high levels of anxiety into our behaviour and our stance to the world became significantly different to all other species.

It is this combination of motor-control and defences to control, powered by the urgency of maintaining safety, that led to the development of speech, language and social contracts. Speech is an ideal means by which the controller issues the controls and for the controlee to try to negotiate, ameliorate and clarify those controls.

Loss of autonomy and the use of top-down thinking are antithetical to equanimity and enlightenment and so the hominins became excluded from the experience of paradise, an experience that is fundamental to the equanimity of the animal mind. I have thought hard about the use of the word 'paradise', for it has collected a lot of clutter over the years, but it is a term the early thinkers used in their attempts to describe an unfamiliar, desirable mental state and I have continued that tradition. I suggest that paradise is a timeless harmony that is experienced when individual behaviour stems solely from bottom-up autonomous activity (the experience of all animals). It describes the freedom experienced by the individual from having no top-down responsibility for one's own responses and it is experienced when one is free from heteronomous controls. This is a desirable state that could be restored to us in a time to come.

Understanding paradise hinges upon determining the difference between bottom-up and top-down activity. You cannot have top-down control, with speech and language, without the loss of paradise, they are mutually exclusive. In the future we will have to decide whether we will promote and encourage acts of motor-control thereby maintaining a loss of autonomy across the species, or to eliminate motor-control, allowing autonomy and a bottom-up mental organization to return. The freedom that comes from not having to comply with heteronomous controls would be life-changing for it would allow the individual to bask in the ease of paradise – a sure sign that autonomy has been restored.

For the first time the underlying quest of all the major religions for enlightenment can be understood as having a basis in scientific truth. We should not be seeking man-made gods in fictitious heavens; instead we should be seeking the real-life experience of animal autonomy with its associated feelings of equanimity, freedom and bliss, and this should be the norm for everybody throughout their life on earth. The experience of paradise follows the restoration of autonomy as surely as anxiety and unease follow the experience of motor-control.

Our species has been excluded from the timeless harmony of paradise that animals inhabit, hence our erratic and incomprehensible flirtations with God and religion. In the process of being expelled from paradise we were blinded to its existence and to the reason why we were expelled, so we are lost with little or no knowledge of how to return. In the past we have been dependent on poets, philosophers and thinkers to explore this puzzling area, hoping that each time they wrote, it would increase our understanding, but the factual details of paradise have remained elusive. This time it is different because the biological reality of loss of autonomy, motor-control, heteronomy and loss of anoestrus can now be understood as the biological elements that led to our expulsion from paradise. These are tangible scientific facts and when they are examined carefully they reveal the way in which autonomy and equanimity were lost and show how the heteronomous barriers to paradise were erected and maintained.

All animals lead independent lives and their autonomous bottom-up behaviour is non-taxing to them in the sense that their conscious mind is not responsible for the actions of its body and therefore cannot be answerable to any other animal (or deity) for its conduct. This creates a mental stance that is quite different from the human's stance to the world, for it is free from criticism, reassessment and heteronomous control. The conscious mind of the animal is not responsible for initiating behaviour, it is not responsible for the outcome of that behaviour, nor does it decide if one course of action is better than another (those decisions are built into the species' safety/danger scales), which means that the main role of the conscious mind is simply that of a non-judgemental observer. That is why the animal cannot be imputed for its behaviour (the lion cannot be criticised for killing the impala, nor the beaver for felling the trees).

Hence the conscious mind of the animal has no thoughts or anxieties about what it should or should not do, for that is solely the function of the safety/danger scales. The conscious mind of an individual does not create, maintain or change any of these response programmes, indeed it is free from any responsibility so to do. Any adjustments to the safety/danger scales are made on a completely different timescale through the non-conscious selection processes of evolution. The animal relies on the fact (without being aware of the fact) that the responses available to it are the best possible because they have survived every selection pressure that the individual's ancestors had experienced to date.

Evolution flows inexorably, 'selecting' those individuals which survive. There is no organizer, or deity, thinking or planning an outcome or working out which is the best behaviour to enact. There is no responsibility upon the individual, nothing for the individual to do except enact its own inherited bottom-up responses as they gain priority on the safety/danger scales. There is nothing here that requires self-conscious decision-making or the need to worry about the responses or any need to improve those responses. This lack of responsibility for one's actions is a feature all animals have in common and it means that their mental stance to the world is comfortably light and carefree (certainly not weighed down with the burdens of heteronomous control).

As you travel around the world and see the artefacts of civilization and materialism, the cities, roads, vehicles, technology, hospitals, law courts, prisons, schools, offices, factories, shops and theatres, as well as the areas of conflict and warfare, it is hard to believe that it has all arisen because we have opposable thumbs strong enough to grip, hold and control our fellow beings. If we had had paws, or pads, or smaller, weaker thumbs, then we would have remained as animals. By suffering the indignities and injustice inherent in motor-control, the hominins lost their autonomy and the open, natural, free flow of their bottom-up animal behaviour became restricted, stilted, and contrived by top-down heteronomous directions. Hominins became set against each other and in having to find safety from each other we have created entirely new behavioural patterns to the extent that we are no longer seen as animals.

We are now approaching a turning point and this book contains the first outline of a very different evolutionary explanation of why we behave in the way that we do. At some stage in our development we will have to think through and overcome our human exceptionalism and see ourselves as the uneasy, troubled creatures that we have become since we parted from the apes. We need to see that we are the only animal that has to defend itself from heteronomous control and to see that we have lost our original equanimous animal mind and replaced it with a state of permanent unease. Survival on these terms has come at a high price for it has meant that, as a species, we are mired in a very stressful place having to deal with heteronomous relationships that no animal has ever experienced and in so doing we have lost the animal's mental ability to function bottom-up. We view human nature from an anomalous position without realising just how anomalous we have become.

Many would claim that *Homo sapiens* have broken free from the 'bestiality' of animal behaviour, particularly so for the female who no longer comes into oestrus (on heat), that is, she is not tied to the hormonal cycles that determine when the animal female 'stands' to mate. At present this is thought to be some sort of liberation, however, if all factors are considered, the opposite is true, for the lack of an overt oestrus is a bodily recognition that the female cannot defend herself from male coercion throughout any part of her cycle. The human female (without assistance) is unable to maintain the safety from sexual penetration that is the essence of the animal's long periods of anoestrus. Her loss of an overt oestrus is a result of being overwhelmed, it is the 'best' she can achieve. This is not a victory for civilization, dignity, decorum, decency, marriage or intelligence; she is simply adjusting to a tragically difficult situation.

Overt oestrus and anoestrus have been lost to the human female due solely to the power of the male hand that is able to grip and enforce coercive mating. The opposable thumb is a very small anatomical difference but, in combination with terrestriality, it inflicted such a catastrophic loss of autonomy that it changed the course of hominin evolution.

Changes brought about by evolution cannot be called 'errors' or 'miss-moves' because there is nothing within the operation of evolution to accept responsibility for outcome. However, we humans, who still suffer the tragic consequences of motor-control, have a case to feel aggrieved. It can be thought that nature (God) has 'erred' in an unprecedented way and we are still seeking to understand how this could have occurred.

It was our vulnerability to being caught and held that led to our 'Fall from Paradise', for once we were subject to heteronomous controls we lost the grace that all animals have as they act freely and autonomously all of time. Human behaviour is awkward, tenuous, premeditated, planned, inhibited, delayed and stilted under the weight of heteronomous controls. We are disingenuous and duplicitous, we are hopelessly corrupt, we struggle, strive and work and we have no rest or peace. We have been excluded from Eden by our ability to motor-control and our subsequent loss of our bottom-up mental organization, and so we no longer experience the bliss and equanimity of the animal mind. We have become enemies to ourselves, preoccupied with gaining deliberate advantage over our fellow beings.

It is hard to overestimate the consequences of this loss of grace; paradise has become so far removed from human experience that we find it almost impossible to comprehend. Bondage, sexual coercion and loss of autonomy have become the norm, and this has created unprecedented behavioural problems that animals have been fortunate to avoid.

This book presents a radical reconceptualization of human behaviour; it suggests a thoroughly ordinary, biological explanation for the mysteries of hominin evolution and answers the question of why we are so different from all other animals. It explains (in terms of natural science) how we lost our autonomy and fell into bondage – how our hands have become our Achilles' heel.

As these principles are verified they will become the backdrop of human existence moulding human behaviour in ways that are undreamt of at present, bringing the brutal battle between autonomy and heteronomy to an end. For not until the power of heteronomous control is eroded and autonomy regains its free expression will we be able to live, like the wild animals, in a state of bliss and equanimity within the gift of nature.